D0772494

DESERTS ON THE MARCH

PAUL B. SEARS

DESERTS ON THE MARCH

Fourth Edition

University of Oklahoma Press : Norman

BOOKS BY PAUL B. SEARS

Deserts on the March (Norman, 1935, 1947, 1980)
This is Our World (Norman, 1937)
Who Are These Americans (New York, 1939)
Life and Environment (New York, 1939)
Charles Darwin (New York, 1950)
The Living Landscape (New York, 1966)
Lands Beyond the Forest (New York, 1969)
(with Marion Rombauer Becker and Frances Jones Poetker)
  Wild Wealth (New York, 1971)

LIBRARY OF CONGRESS CATALOGING IN PUBLICATION DATA

Sears, Paul Bigelow, 1891–
  Deserts on the march.

  1. Man—Influence on nature—United States.
2. Desertification—United States.   3. Agricultural
ecology—United States.   4. Conservation of natural
resources—United States.   I. Title.
GF503.S42 1980   333.7′2′0973   79-6721

Contents

Preface to the Fourth Edition

SINCE 1935 WHEN THE FIRST EDITION OF *Deserts on the March* was published, a short shelf of books dealing with man's impact on environment and its reaction has expanded into a library. Lively concern is replacing public indifference, at times to the point where understanding and judgment have lagged behind zeal. Why then presume to retouch a message that has gathered volume during more than four decades?

Time has taken care of a guess of some demographers that the population of the United States would stabilize at around 160 million by 1960. We are now convulsively aware that the supply of fossil fuel is finite—a warning repeated two years later along with the condemnation of war as a most greedy destroyer of natural resources in a second book, *This Is Our World* (Oklahoma, 1937). Meanwhile, interest in planning, production of food, and the protection of nature has advanced.

Yet human numbers continue to increase faster than the means of survival, to say nothing of decent living conditions. Land is being taken away from the production of food and fiber at around a million acres a year in the United States alone, while world wide the man-made deserts continue to grow by an annual area approaching that of Maine or South Carolina.

Mass production, better designed to consume than to recycle materials and largely dependent upon diminishing supplies of fossil fuel, is, with its prize exemplar the automobile, a major source of pollution. Along with much that is essential in modern life, it produces much that is by no measure indispensable and must be marketed by costly artifice. Applied to agriculture and education, the techniques of mass production are a long-range threat to soil structure and an obstacle to the personal relationship needed for the best teaching and learning.

Clearly this is no time to relax effort, no matter how imperfect and stumbling, nor to abandon reminders that our destiny is bound to that of the planet where our very existence is the result of the events of some five billion years.

It is impossible to acknowledge the many persons to whom I am indebted for whatever of value follows this brief foreword, as it is to offer any ex cathedra solutions. Instead, from the rich variety now available, I venture to suggest the worth of the following:

The Next Hundred Years, by Harrison Brown, James B. Bonner, and John Weir, for the acknowledgment by distinguished scientists that the more elaborate a system of technology, the more vulnerable it is to the slightest failure; and that throughout history the most viable form of human culture has been the simple agrarian.

From the High Plains, by John Fischer, for its vivid

account of the sequence of exploitation in semi-arid North America—mining for flint by Indians and (far more serious) depletion by white men of buffalo, grass, soil, oil, and finally, water.

The superbly-written report to the United Nations, *Only One Earth,* by Barbara Ward and Rene DuBose, not only for its content but also for recognizing that our hope for survival in all our prized diversity rests upon our "ability to achieve an ultimate loyalty to "our single beautiful, and vulnerable Earth."

Taos, New Mexico   PAUL B. SEARS
January, 1980

DESERTS ON THE MARCH

Introducing himself as we were waiting for transportation, a young German told us that he was just finishing a tour of the United States. Asked what had impressed him most, he instantly replied, "Waste, enormous and tragic waste."
(April, 1979)

1
Man, Maker of Wilderness

THE FACE OF THE EARTH is a graveyard, and so it has always been. To earth each living thing restores when it dies that which has been borrowed to give form and substance to its brief day in the sun. From earth, in due course, each new living being receives back again a loan of that which sustains life. What is lent by earth has been used by countless generations of plants and animals now dead, and will be required by countless others in the future. The supply of an element such as phosphorus is so limited that if it were not being constantly returned to the soil, a single century would be sufficient to produce a disastrous reduction in the amount of life. No plant or animal, nor any sort of either, can establish permanent right of possession to the materials that compose its physical body.

Left to herself, nature manages these loans and redemptions in not unkindly fashion. She maintains a balance that will permit the briefest time to elapse between burial and renewal. The turnover of material for new generations to use is steady and regular. Wind and water, those twin sextons, do their work gently. Each type of plant and animal, as far as it is fit, has its segment of activity and can bring forth its own kind to the limit of subsistence. The red rule of tooth and claw is less harsh in fact than in seeming. There is a balance in undisturbed nature between food and feeder, hunter and prey, so that the resources of the earth are never idle. Some plants or animals may seem to dominate the rest, but they do so only so long as the general balance is maintained. The whole world of living things exists as a series of communities whose order and permanence shame all but the most successful of human enterprises.

It is into such an ordered world of nature that primitive man fits as a part. A primitive family living by the chase and gathering wild plants requires a space of ten to fifty square miles for subsistence. If neighbors press too close, the tomahawk of tribal warfare offers a rude but perhaps merciful substitute for starvation. Man in such a stage takes what he can get on fairly even terms with the rest of nature. Wind and water may strike fear to his heart and even wreak disaster upon him, but on the whole their violence is tempered. The forces of nature expend themselves beneficently upon the highly developed and well-balanced forests, grasslands, even desert. To the greatest possible extent the surface consists of mellow, absorbent soil, anchored and protected by living plants—a system buffered against the caprice of the elements, although of course subject to slow and orderly change. Bare ground left by the plow will have as much soil washed off in ten years as the unbroken prairie will lose in four thousand.

Even so, soil in the prairie will be forming as fast as, or faster than, it is lost.

Living in such a setting, man knows little or nothing of nature's laws, yet conforms to them with the perfection over which he has no more choice than the oaks and palms, the cats and reptiles around him. Gradually, however, and with many halting steps, man has learned enough about the immutable laws of cause and effect so that with tools, domestic animals, and crops he can speed up the process of nature tremendously along certain lines. The rich Nile Valley can be made to support not one but more than one thousand people per square mile, as it does today. Cultures develop, cities and commerce flourish, hunger and fear dwindle as progress and conquest of nature expand. Unhappily, nature is not so easily thwarted. The old problems of population pressure and tribal warfare appear in newer and more horrible guise, with whole nations trained for slaughter. And back of it all lies the fact that man has upset the balance under which wind and water were beneficent agents of construction, releasing them as twin demons that carve the soil from beneath his feet, to hasten the decay and burial of his handiwork.

Nature is not to be conquered save on her own terms. She is not conciliated by cleverness or industry in devising means to defeat the operation of one of her laws through the workings of another. She is a very businesslike old lady who plays no favorites. Man is welcome to outnumber and dominate the other forms of life, provided he can maintain order among the relentless forces whose balanced operation he has disturbed.

But this hard condition is one that, to date, he has scarcely met. His own past is full of clear and somber warnings—vanished civilizations buried, like dead flies in lacquer, beneath their own dust and mud.

For man, who fancies himself its conqueror, is at once the maker and victim of the wilderness. Even the dense and hostile jungles of the tropics are often the work of his hands. The virgin forest of the tropics, as of other climes, is no thicket of scrub and thorn, but a cathedral of massive, well-spaced giant trees under whose dense canopy the alien and tangled rabble of the jungle does not thrive. Order and permanence are here—these giants bring forth young after their own kind, but only as fast as death and decay break the solid ranks of the elders. Let man clear these virgin forests, even convert them into fields, he can scarcely keep them. Nature claims them again, and her advance guards are the scrambled barriers through which man must chop his way.

In the early centuries of the present era, while the Roman Empire was cracking to pieces, the Mayas built great centers in Central America. Their huge pyramids, massive masonry, and elaborate carvings are proof of capacity and leisure. They also indicate that the people who built them probably felt a sense of security, permanence, and accomplishment as solid as our own. To them the end of their world was no doubt unthinkable, except as a device of priestly dialectic or an exercise of the romantic imagination. Food there was in abundance, furnished by the maize, cacao, beans, and a host of other plants of which southern Mexico is the first home. Fields were easily cleared by girdling trees with sharp stone hatchets. You can write your name on plate glass with their little jadeite chisels. The dead trees were then, as they are today, destroyed by fire, and crops were planted in their ashes.

Yet by A.D. 900 all of this was abandoned and the Second Empire established northward in Yucatan, to last with varying fortunes until the Spanish conquest. Pyramids and stonework then became the playground of the jungle, so hidden and bound beneath its knotted mesh that painful labor has been required to reveal what is below. Surviving in Yucatan in humble villages are the modern people, unable to read the hieroglyphs of their ancestors, and treasuring only fragments of the ancient lore that have survived by word of mouth. There persists among these people, for example, a considerable body of knowledge concerning medicinal plants, their properties and mode of use. But the power and glory of the cities is gone and in their place, only ruins and wilderness. Their world, once so certain, stable, dependable, and definite, is gone. And why?

Here, of course, is a first-rate mystery for modern skill and knowledge to unravel. The people were not exterminated, nor were their cities taken over by an enemy. Plagues may cause temporary migrations, but not the permanent abandonment of established and prosperous centers. The present population to the north has its share of debilitating infections, but its ancestors were not too weak or wasted to establish the Second Empire after they left the First. Did the climate in the abandoned cities become so much more humid that the invasion of dense tropical vegetation could not be arrested, while fungi, pests, insects, and diseases took increasing toll? This is hard to prove.

Were the inhabitants starved out because they had no steel tools or draft animals to break the heavy sod that formed over their resting fields? Many experts think so.

Certainly the soil of the wet tropics is very different from the deep, rich black soil of the prairies. Just as soaking removes salt from a dried mackeral, so abundant rainfall leaches the nourishing minerals from these soils. In the steaming hot climate the plant and animal materials that fall upon the ground rot quickly, sending gases into the air and losing much of what is left in the pounding, soaking wash of the heavy tropical rains. Such organic material as may remain is well incinerated when the forest covering is killed and burned, as it was by the ancient Mayas, and still is by their descendants. Such a clearing will yield a heavy crop for a few seasons, by virtue of the nutrients in the ashes and what little is left in the soil. Presently the yield must decline to the point where cultivation is no longer possible. A fresh clearing is made and the old one abandoned. Step by step the cultivation proceeds farther from the place of beginning. Whether the idle fields, forming an ever-widening border around the great cities, came to be hidden beneath an armor of impenetrable turf or completely ruined by sheet erosion and puddling is immaterial. The restoration of fertility by idleness has proved a failure even in temperate climates. It is not a matter of one or even several human generations, but a process of centuries. The cities of the Mayas were doomed by the very system that gave them birth. Man's conquest of nature was an illusion, however brilliant. Like China before the Manchu invaders, or Russia in the face of Napoleon, the jungle seemed to yield and recede before the

Mayas, only to turn with deadly, relentless deliberation and strangle them.

So much for an example of failure in the New World. How about the Old—the cradle of humanity? Here there are striking demonstrations of apparent success, long continued, such as one finds in eastern China and the Nile Valley. On the other hand are many instances of self-destruction as dramatic as that of the Mayas—for example, the buried cities of the Sumerian desert. Let us examine both failure and seeming success; after we have done so, we shall realize how closely they are interwoven.

The invention of flocks and herds of domestic animals enabled man to increase and prevail throughout the great grassy and even the desert interior of the Old World. Food and wealth could be moved on the hoof. A rough and ready "cowpuncher" psychology was developed as a matter of course, combining a certain ruthless capacity for quick action along with an aversion to sustained and methodical labor, except for women. Living as these people did, in a region where water was none too abundant and pasture not always uniform, movement was necessary. Normally, this was a seasonal migration, a round trip like that of the buffalo and other wild grazing animals. But from time to time the combination of events brought about complete and extensive shifts.

Where moisture was more abundant, either directly from rain or indirectly through huge rivers, another invention took place. This second invention was the cultivation of certain nutritious grasses with unusually large fruits: the cereals. Probably not far from the mouth of the Yangtze River in southeastern China, rice was domesticated, as were wheat and barley, at the eastern end of the Mediterranean, both in Iraq (Mesopotamia) and Egypt. Along with these cereals many other plants, such as beans, clover, alfalfa, and onions were grown. This invention provided food cheaply and on a hitherto unprecedented scale. Domestic animals could now be penned, using their energy to make flesh and milk instead of running it off in continual movement to obtain grass and water. Other animals, such as the cat and dog, assisted man in guarding his stored wealth against the raids of rats and robbers. Such large animals as the ox and ass saved him the labor of carriage and helped in threshing and tillage. The people themselves became accustomed to methodical and prolonged labor. They devised means of storage and transport, and they developed commerce. Mechanical contrivances proved useful and were encouraged. On the other hand, such folk were not celebrated for their aggressiveness or for an itching foot. As they became organized and accumulated a surplus of skill and energy, they developed great cities and other public works, with many adornments.

The history of early civilization can be written largely in terms of these two great inventions in living: the pastoral life of the dry interior and the settled agriculture of the well-watered regions. Their commerce, warfare, and the eventual, if imperfect, combination of the two make the western Europe of today. What of their effects upon the land?

Wherever we turn, to Asia, Europe, or Africa, we shall find the same story repeated with an almost mechanical regularity. With few exceptions, the productiveness of the land has decreased, fertility has diminished, and soil has been destroyed at a rate far in excess of the capacity of either man or nature to replace it. The glorious achievements of civilization have been built on borrowed capital to a scale undreamed by the most extravagant of monarchs. And unlike the bonds that statesmen so blithely issue to—and against—their own people, an obligation has piled up that cannot be repudiated by the stroke of any man's pen.

Uniformly the nomads of the interior have crowded their great ranges to the limit. We shall see later what a subtle matter this crowding can be—the fields may look as green as ever, until the inevitable drier years come along. Then the soil becomes exposed, to be blown away by wind, or washed into great flooded rivers during the infrequent, usually torrential rains. The cycle of erosion gains momentum, at times conveying wealth to the farmer downstream in the form of rich black soil, but quite as often destroying and burying his means of livelihood beneath a coat of sterile mud.

The reduction of pasture, even with the return of better years, dislocates the scheme of things for the owners of flocks and herds. Raids, mass migrations, discouraged and feeble attempts at agriculture, or, rarely, the development of irrigation and dry farming result—and history is made.

Meanwhile, in the more densely settled regions of cereal farming, population pressure demands that every resource maintain yield. As long as rich mud is brought downstream in thin layers at regular intervals, the valleys yield good returns at the expense of the continental interior. But such imperial gifts are hard to control, increasingly so as occupation and overgrazing upstream develop. In the course of events farming spreads from the valley to the upland. The forests of the upland are stripped, both for their own product and for the sake of the ground that they occupy. Growing cities need lumber and fuel, as well as food. For a time these upland forest soils of the moister regions produce good crops, but gradually they too are exhausted. Imperceptibly, sheet erosion moves them into the valleys, with only temporary value to the latter. Soon the rich black valley soil is overlaid by pale and unproductive material from the uplands. The uplands then may become an abandoned range of gullies, or in rarer cases human resourcefulness may come to the fore and by costly engineering works combined with agronomic skill defer the final tragedy of abandonment.

Thus have we sketched, in broad strokes to be sure, the story of man's destruction upon the face of his own Mother Earth. The story on older continents has been a matter of millennia, as we shall see. In North America it has been a matter of not more than three centuries at most—generally a matter of decades. Mechanical invention plus exuberant vitality have accomplished the occupation of a continent with unparalleled speed, but in doing so have broken the gentle grip wherein nature holds and controls the forces that serve when restrained, destroy when unleashed.

2
The Wisdom
of the Ages

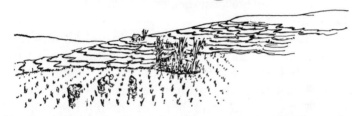

Is THE BATTLE really a losing one to date? Have not invention, energy, and discipline consolidated the gains of mankind securely against all danger, excepting our own selfishness and capacity for mutual destruction in time of war and peace? What of the wisdom of the East? What of the vast and varied land known as China, containing about 22 percent of the world's population on slightly more than 6 percent of the world's total area? What of Mother India, with close to 10 percent of all living people on little more than 2 percent of our planetary surface? What of the fertile Valley of the Nile, hedged by desert, crowded since ancient times, and struggling today to maintain a viable economy? Is not the fact that these venerable communities have survived the millennia sufficient proof that, even without science as we know it, man's ingenuity is match for the vicissitudes of an impersonal, dynamic, and finite environment? How much better mankind's chances now that we are beginning to understand the rules of the game?

To gauge those chances fairly, we would do well not to join those whom P. B. Medawar calls "the semi-literates who regard any work done earlier than in the past year or two as merely of antiquarian interest." For the whole human adventure, indeed that of life itself, is one process; what has happened, and is happening must either enhance or diminish the options of the future.

The Chinese civilization has outlasted all others of the ancient and modern world. Somewhat larger than the United States, China is the home of more than three times as many human beings. Its arts, inventions, and crafts are vouchers for the intelligence and versatility of its people. To China the western world is indebted for movable type, fireworks, paper, silk, and possibly the mariners' compass. Its traditional products include exquisite ceramics, lacquer, carving, and not least, cereals, fruits, and other vegetable materials.

Farming and gardening in China probably reached the utmost in efficiency possible without the aid of modern science. The investment in human energy per acre was great, yet measured no more than that employed by agribusiness using fossil fuel. The return of organic wastes to the soil was as complete as physically possible. Communication and transport throughout the empire ensured the introduction and testing of crops and their improvement through selection. From China have come rice, soybeans, and alfalfa. The wild mountain forests of its south have given us the apple.

Capping these resources and practices has been something basic to the folkways of any culture. That something is the intangible business known as sanc-

tion, or accepted value. Tillage and its products, notably rice, were celebrated in imperial ritual. And in
principle—however the individual farm laborer may
have been exploited—the farmer was honored above
merchant and soldier and only below the scholar.

It has been said, and probably with truth, that the
land of southeastern China is almost unique in bringing forth as much today as it ever has. Its people are
industrious, frugal, and intelligent. If they were here
in competition with our own farmers, it is likely that
they would put all but the best of them out of business.
These being the facts, what of China?

China today is undergoing profound political, social,
and economic changes that are especially concerned
with the use and management of environment. But
before we consider this situation, we need to look at the
record, for the present is child of the past. How has this
great Oriental empire fared through centuries of
skilled and thrifty farming?

To begin with, its millions have never, through
recorded history, been secure from the threat of starvation. Flood and famine, frequently local but widespread at intervals, have been remorseless in their re-
currence. When seven thousand Chinese Communists
were wiped out in four days, the government then in
power reminded its critics, truly enough, that this particular project in "political sanitation" was a small
matter beside the deaths in a single provincial famine.
The compliment, if that is what it was, later was returned with interest.

During the centuries every available bit of the eastern Chinese plain, once heavily forested, had been
cleared for fuel, building material, and to increase the

amount of tillable land. This land in turn has been terraced and contoured with geometrical perfection. Its farmers have saved every scrap of garbage and other organic matter. They trudged long distances to the cities and villages with containers balanced over their shoulders to bring back to the land what writers on agriculture call by the euphemistic name of "night soil"—human excrement. All of this, together with the rich black ooze left on flood plains by high water, has been worked into their tiny fields with the pains bestowed by a Dutch housewife on her window garden. Somehow too the Chinese farmer has learned what the Romans knew and their successors forgot—the value of legumes in enriching the soil. Unlike medieval Europeans, he has consistently practiced legume rotation with its added advantage of giving him the soybean from which Chinese cooks have evolved an impressive array of edibles.

The late John M. Coulter, who developed one of the world's great botany departments at the then youthful University of Chicago, had been born in China and in 1922 was planning to use his retirement to advise the Chinese on agriculture. Asked why, in view of the remarkable quality of traditional practice, he said that even so, modern genetics, soil science, and plant physiology were much needed. Yet the general standard of practice was so far above that in many parts of the western world as to admit of no comparison. One of the finest achievements of the Chinese farmer has been the conversion of the Red Basin of Szechwan from an incipient bad land, supporting fewer than 145,000 people in 1710, into a flourishing, beautifully terraced agricultural countryside of more than 70 million in 1957.

Yet nothing is more in error than to think that China, by that year, had become in any real sense self-sustaining. Its productive area has been nourished, not only by frugal recycling, but by the contributions of its great rivers, brought down from their sources in the vast, sparsely peopled mountains. These streams are fed by inexhaustible snows and bear mineral nutrients along with organic material produced by the lush plant cover of the world's largest mountains. Should modern industry penetrate to these headwaters, as it has to our own Rocky Mountains, downstream disaster would follow.

As we shall see, there is hope, if not assurance, that this consequence may be averted. But in the past the persistence of the protective living cover of these upper watersheds has been more a matter of happy chance than of deliberate policy. Had means existed for the ready transport and marketing of their timber, they would doubtless have been stripped long ago. Even so, as with wooded mountains throughout the world, the pressure of population has continuously surged against the Himalayas.

Some decades ago, potatoes were introduced into western China by missionary priests who hoped to increase insurance against recurring famine. At first the Chinese, repeating the experience of western Europe, regarded the potato with contempt as inferior food, fit only for those who could do no better for themselves. But when it was discovered that this descendant of Andean ancestors could grow at elevations where many of the staple food plants could not, there was a rush to strip the foothill forests and replace them with potato patches.

Eventually—and not too long afterward—an inexorable principle of biology came into play. Stability depends upon variety. As the Irish so tragically discovered in the 1840's, there is grave peril in relying upon a single source of food. Without other crops to furnish reserves when the potato fields were blighted, the land so recently put to use had to be abandoned. The tidal wave of humanity receded and, we are told, the forest returned. But should the attack on the native plant cover continue, not even the favorable climate could produce enough growth to prevent serious erosion or protect the lower Yangtze against destructive flood.

Unlike the Yangtze, the Hwang (Yellow River) has been notorious for its disastrous floods. Starting like the Yangtze in the snowy, wooded, sparsely peopled mountains of Tibet, it flows down through a drier, more continental region than its southern fellow. Much of its upper valley is plateau covered with the fine windblown sediment known as loess, in places to the incredible depth of several hundred feet. As in our own western high plains, the natural cover is grassland, but the pressure of population has been insistent. Wherever possible, cultivation has been attempted; elsewhere, herd and flock have taxed the pastures to the limit.

Fertile when supplied with water, vulnerable to erosion if bare of plant life, the denuded loess is carved into fantastic badlands of vertical cliffs by floods so frequent and intense that the stream courses shift often. Downstream the sediment deposit is heavy and dikes have raised the river bed above the surrounding plain, greatly increasing the risk to life and property. Adjacent to but outside the true climatic Asian desert, the flood plain of this "River of Sorrows" has, time out of mind, witnessed the somber tragedy of man's struggle against cultural desert of his own creation.

Prudent land management has been official Chinese policy since the reign of Emperor Shen-nung in 2700 B.C. This, combined with the most frugal husbandry, has assured survival despite invasions and other environmental disasters. But if the record proves anything, it is that the best of traditional measures are powerless to prevent famine, disease, poverty, and environmental damage unless human numbers are effectively controlled.

The present government of China is committed, with the aid of modern science and powerful political control, to the abolishment of ancient evils. Credible reports indicate remarkable progress in reforestation, and in controlling hunger, disease, unemployment, and even population increase. But we are reminded that the success of these measures rests with decisions made at the center of power, to which we must add the impact of consumptive modern technology and international developments, along with the enormous inertia of ancient ways.

So much for China. What of India, whose teeming millions flit past the occidental imagination in a kaleidescopic mixture of splendor and wretchedness? In montage we may see heaps of jewels, plump elephants, forests of precious woods, groves of spice and tea, rice fields of cane and rice. But always there are flashes of hungry faces, ragged bodies, crowding beggars, and all that symbolizes misery. No matter how many rotund and prosperous Indians one may have met face to face, the inescapable image of India to most western minds is that of a gaunt and hopeless figure, standing on sunbaked, barren clay beside an undernourished, wizened cow. Lest this seem too gross a caricature, let an Indian, Aga Khan III, speak:

The ill-clad villagers, men, women, and children, thin and weakly, and made old beyond their years by a life of under-feeding and overwork, have been astir before daybreak, and have partaken of a scanty meal consisting of some kind or other of cold porridge, of course without sugar or milk. With bare and hardened feet they reach their fields and im-mediately begin to furrow the soil with their lean cattle of a poor and hybrid breed, usually sterile and milkless. A short rest at midday, and a handful of dried corn or beans for food, is followed by a continuance till dusk of the laborious scratching of the soil. Then the weary way homewards in the chilly evening, every member of the family shaking with malaria and fatigue. A drink of water, probably contami-nated, the munching of a piece of hard black or green cham-patire, a little gossip round the peepul [pipal] tree, then the day ends with heavy unrefreshing sleep, in dwellings so in-sanitary that no decent European farmer would house his cattle in them.

The specter of famine is never far away in India. It is estimated that 10 million people died in the famine of 1770 amid scenes of suffering so harrowing that only a morbid mind would dwell upon them. So slight was the margin of supply that even three successive years of good crops thereafter could not restore the balance. Not enough people were left to work the fields. In 1865 one-third of the population of Orissa died by famine, and in the subsequent three years 1.5 million victims were claimed. The roster of Indian famines reads with an appalling monotony.

India is smaller than greater China, but compares roughly with China proper in area and population. Northward, the humid Gangetic plain and the arid Punjab are watered, as is China, by the huge Himalayas. Hence they are mostly covered by rich material washed over their surface. In the same sense as China, they are dependent on treasures stored up in the past, as well as upon water and fertility brought today from the thinly populated Tibetan regions beyond their borders. They enjoy the added protection that comes from very moderate agricultural exploitation of their northern boundaries. Here considerable areas are occupied by tribesmen whose farming operations have been casual in nature. More or less on the move, they have allowed their fields to grow back into forests after a few years of use—a custom that the climate permits. Thus the protecting girdle of vegetation is maintained from Tibet south into the hills of India. Should this be methodically stripped and the land put to the plow, it is easy to see that great skill would be required to prevent disaster to the more populous, lower valley regions.

The Dekkan, or peninsula of India, lying south of the regions named, is separated from them by a line of low mountains from which it receives drainage. With the exception of its west coast, it is immediately dependent upon rainfall, yet supports a dense population. The soil is peculiarly subject to erosion, and this evil has been intensified by removal of the forest cover from the hills that form the watersheds. Modern industrialism has encouraged and hastened this process. While the soil is varied, the direct cause of famines has been largely a recurring lack of rainfall. Soil management has not been on a level with that in China. Until recently, fertility seems to have been maintained by a process of growing very little more than was needed to sustain life and consuming it locally. In the absence of means of ready transportation, even a local crop failure could produce starvation, and often did just that. Under British rule many conditions were alleviated, but there was also a vast increase in population. Today, there is still hunger in the event of local crop failures. Food in ample quantities can be brought in, but must be paid for when it comes. Furthermore, the very railroads that facilitate the transportation of food have hastened the clearing of the forests that had earlier helped to check erosion.

Long before Indian independence in 1947, scientists and civil servants, British and Indian, did what they could to bring about better land use and management, including irrigation. Some authorities believe that India could more than produce what it needs, even though they express concern as to ways, means, and ultimate effect on the soil. The truth is that this ancient land exemplifies both the benefits and the costs of cultural inertia. Rural village life, despite hardships and disasters, is far less vulnerable than more highly urbanized, undustrial economies. But institutions such as child marriage and protection of undernourished cattle regardless of quality and utility intensify the bondage of human numbers. Whatever violence these, with other practices and beliefs, may do to western ideas, they have had the force of logic in the culture of this ancient land.

Even with the hopes aroused by independence, there is no assurance of a healthy, self-sustaining economy in the near future. The more fortunate parts of India are in debt for what they receive from beyond their own borders. As to the rest, the price of survival has been too largely a state of suspended animation, low vitality, and the constant threat of hunger for millions whose needs so greatly exceed the present capacity of environment to meet them. Thus far, who has the upper hand here, man or nature?

But let us move on to Egypt, that half-oriental teacher of the western world. Less than 4 percent of an area larger than France and Germany combined is habitable, and here the population is about two thousand per square mile. With negligible rainfall, Egypt has by no means been self-contained; moisture and silt from the sources of the Nile have enabled her to exist by methods brought down from the days of the Pharaohs.

Laborious and primitive, this mode of agriculture, or rather gardening, was not without merit. Two American scientists, Homer Shantz, of Arizona, and Leo Melchers, of Kansas, came to this conclusion quite independently. Shantz, invited to examine the agriculture of Ethiopia, was at first impressed with the improvement that would result if modern machinery were to replace the wooden plows, drawn by ox or ass, that merely scratched the surface. Wisely delaying his recommendation to "modernize," he realized that this African country still had its soil after two thousand years, while his own had lost vast amounts through erosion. He counseled no immediate radical change.

This experience was almost exactly repeated by Melchers, in Egypt at the request of its government and quite uninformed as to the earlier visit of Shantz to Ethiopia. But even before Egypt became independent of Great Britain in 1953, the pressure to change old patterns was becoming irresistible. Primitive methods had conserved the nutrients brought down by the annual bath of mud, administered with gentle dignity by the stately Nile. The water that brought it was impounded and fed out by ancient but clever and effective measures along the course of that river.

In the delta at its mouth, however, where there are no high levels at which the water could be stored, only one crop could be grown each year, in soil still moist from the flood. Knowing that the climate would permit crops at any time of year if water were available, the authorities, in true modern high-pressure style, worked out a system of ditches to hold water in storage. By this means three crops a year could be grown in the delta and the ground kept constantly at work. To an overpopulated country, heavily dependent on exports of high-quality cotton and staples, this must have seemed an answer to prayer.

But like so many attempts to circumvent principle by using a law (in this instance of Nature), this action proved self-defeating. The ditches became clogged with silt instead of allowing it to spread as usual. Soon the delta of the great river Nile, richest in the world, began to show symptoms much like those of a broken-down, one-crop cotton or tobacco farm. Compensation in the way of costly fertilizers has been exacted (as was predicted in the original, 1935 edition of this book).

Meanwhile, in further attempts to improve ancient ways by the use of modern technology, the Aswan Dam was completed with Russian aid in 1969. Arresting the northward flow of the Nile, the dam has also withheld the annual bath of nutrient mud and increased the need for artificial fertilizers. And it has had the tragic effect of increasing the prevalence of a dread snail-borne disease. These are serious discounts against the benefits of generous electric power production and controlled irrigation.

Outside of Egypt, what of the rest of northern Africa? The seacoast, once a famous wine district, is now unfit for that purpose. Egypt itself, beyond the immediate borders of the Nile, is desert. Both north and south of the Sudan is evidence of misuse and deterioration—ground cover gone, soil washed and blown. In East Africa the pastures have been so overloaded with stock that here too the soil has been exposed and washed away.

A curious exception occurs in places where the infestation of the accursed tsetse fly has kept down the cattle population. This pest, which has been a source of so much loss and destruction to the cattle industry, actually appears as a blessing when the longtime welfare of the continent is considered. It reminds us of the potato blight in western China, destroying food, it is true, but preventing destructive agriculture that would damage the watersheds supplying the populous districts near the coast. Again, it is like the boll weevil in the southern states that, by ruining the cotton crop, finally forced the farmers to diversify, to their own lasting benefit. Or like the wild hill people who occupy the northern reaches of India with their casual, precarious agriculture that is insufficient to give them more than a meager living, but allows the forest to keep its foothold so that the main part of India suffers less than it otherwise would.

There is not much in the story of China, India, and Egypt to suggest that an entire continent can be exploited with the efficiency of the machine age while its inhabitants multiply and enjoy what the politicians speak of as the "American standard of living."

In short, there is not much in the record of lands occupied by man before he crossed the Bering Strait into America to show that he has been able by hard toil and trial-and-error cleverness to preserve the ability of his environment to support him while increasing his own numbers indefinitely. This is not to say that environmental damage is a simple function of population growth. Vast damage to North America was done by a population less than half what it is now while some of the densely populated parts of western Europe have husbanded their resources well.

Nor should it be thought that application of modern scientific technology need be disastrous, often as it has been. Rather it requires what the late, wise engineer Arthur Morgan called "conclusive engineering analysis," by which he meant that planning dare not stop at the purely technical solution to engineering problems, but should envisage the total environmental, economic, social, and aesthetic consequences of any project under contemplation.

3

Hungry Europe

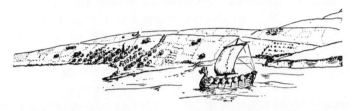

ONE OF THE OLDEST Greek books that survives is a farm almanac. It is a good one, too, the *Works and Days* of Hesiod, dating probably from the eighth century B.C. It is addressed to a public of respectable small farmers, living on their own land and carrying on with the help of one or two slaves. Even today its precepts would be rated as sound and sensible. The course of time is measured by the stars, by the arrival of the swallow, or the fall of the leaf:

Mark, when you hear the voice of the crane who cries year by year from the clouds above, for she gives the signal for plowing, and shows the season of rainy winter; but she vexes the heart of the man who has no oxen. . . . When Zeus has finished sixty wintry days after the solstice, then the star Arcturus leaves the holy stream of Ocean and first rises brilliant at dusk. After him the shrilly wailing daughter of Pandion, the swallow, appears to men when spring is just beginning. Before she comes, prune the vines, for it is best so. But when the House-carrier [snail] climbs up the plants from the earth to escape the Pleiades, then it is no longer the season for digging vineyards, but to whet your sickles and rouse up your slaves. . . .when the artichoke flowers, and the chirping grasshopper sits in a tree and pours down his shrill song continually from under his wings in the season of wearisome heat, then goats are plumpest and wine is sweetest.

Not so precise as our printed calendars, but more to the point and in that sense more accurate. The old farmer of today who starts hunting the delicious morel mushroom when the apple trees burst into bloom comes back with a full basket oftener than his methodical city friend who has noted the date April 1 for his foray. From Hesiod we get the picture of a Greek farming life that provides its own necessities and to spare, and of countrymen not ashamed of their tasks.

Within four centuries of Hesiod's time Plato, Lover of Wisdom, wrote of changes in the Grecian landscape during his lifetime—forests felled, springs dried, and rock bared by the washing away of fertile soil. Quite simply, he had seen what has happened with tragic consistency in many places and throughout history. As wealth and cities grew, owners left the land and became chiefly concerned with what rent they could get from those who worked it. These in turn were less moved by love of place than by what they could squeeze out of the soil with their own labor or, if more prosperous, the labor of their slaves. As holdings became larger and the status of the worker declined, small owners no longer able to compete gave up the struggle and fled to the cities.

As the status of the farm operator declined, so did the quality of husbandry. In the end, of course, Athens became dependent upon shipments of grain from beyond the Dardanelles; when enemies shut off the supply by sea, she was lost. The decline of power is always a complex matter, with explanations that vary with chroniclers; but caught in the pattern, in Greece as elsewhere, was the tragic fact of environmental decline, the effects of which are still evident to visitors in that beautiful and classic land.

As a farmer the modern Italian is one of the best. He can move in on an abandoned New England stone pile and make it pay. But the story is not so simple as it may seem; it is not enough to speak of things as they are. What is now is the outcome of what once was.

We have been told that Cincinnatus left his plow in 468 B.C. to become Roman dictator and returned to it when that task was finished. He exemplified the counsel of Plato, that power should be its own reward, an obligation as much as a privilege. Although he is known to have defended class distinctions, he was no more ashamed to work his own land than Abraham Lincoln was to milk the family cow, until his career made that impossible.

But by the second century B.C. Roman farming had changed from a yeoman to a proprietary enterprise, with controls shifting to the wealthy and the status of the actual worker declining. Marcus Portius Cato the Censor died in 149 B.C., three years before the destruction of Carthage, an event that he had urged, but not before he had written an important treatise, *De Agri Culturu*. Having known Carthage at firsthand, he saw in its economic success a threat to Rome. What else he may have known is an intriguing question, but at any rate it was soon after his death that the senate ordered the extensive writings on agriculture of one Mago of Carthage to be preserved and translated from Punic to Latin.

Cato, like Columella and Virgil in the next century, recommended sound practices whose reasons could not have been known until the days of modern science. A prime example is the use of nitrogen-fixing legumes as a part of crop rotation. But Cato was a man of many interests and much responsibility at a time when wealth and slavery were increasing. Land use was passing from a means of survival for the operator to a source of profit for whoever owned the increasing acreages. The immediate energy required came from the muscles of men and animals and the food that sustained them; the pattern of operation was set by men like Cato and its execution entrusted to agents or overseers.

Inevitably, ownership became exploitation, both of worker and the land. As the status of the worker declined, so did the quality of the land. But even before this happened, the pressure of numbers and the demand for wood (both for construction and fuel) and upland pastures had caused floods, erosion, and swamping of rich lowland plains. The Pontine Marshes and other malarial areas south of Rome were sources of concern even before the Republic gave way to the Empire. Their reclamation under fascism was one of the few good deeds of that regime.

After Rome had passed from yeomanry through tenantry to a pattern of absentee ownership of vast estates, it became no longer self-sufficient and was forced to depend on shipments of foreign grain. Augustus, who combined a good deal of the moralism and practicality that Rome was to bequeath to the Christian church, was concerned at the decay of agrarian life. It may have been no accident that two of the great poets of his day lauded the farmer's simple way of life, Horace with sentiment, Virgil perhaps as an assignment.

But the wise counsel of the Republic was doomed as the Empire expanded beyond its powers of defense and the farmsteads beyond the capacity of responsible ownership. An exception seems to have been the fertile lands of northern Italy and possibly southern France, watered and nourished by the Alps. Here the farming wisdom of the Romans persisted, explaining the high skill of the Italian farmer noted above; his terraced gardens, vineyards, and olive groves, works of perfection, somehow survived the ravages of invaders from the eastern steppes and the gloomy forest lands of northern Europe.

Elsewhere around the Mediterranean decline has been the rule. Much of North Africa, once fertile, is no longer so. Poor management, the ravages of war, and overgrazing by goats have broken whatever power the original soil had to temper the effects of climate. In interior Africa, as well as on other continents, desert margins are expanding, not because of climatic change but because of human activity. Such is the verdict of a multination conference assembled in Kenya in 1978 to consider the current expansion of deserts, not only in Africa but elsewhere. Significant of man's role in this process has been the use of the term "desertification" (the making of deserts) instead of mere "desertization" (development of deserts).

In Spain the Moors were excellent farmers, handling

the land well during the time of their enlightened civilization, fittingly described by an artist friend as "too finely perfumed to last." Making all possible allowance for the lack of sympathetic insight into Spanish culture and history so long prevalent in the English-speaking world, it is clear that grazing—at the expense of woodland and farmland, over which moving herds had right of way—was favored over agriculture. This, the flow of newly discovered gold from the Americas, and what has been called "the tragic padlocking of the Spanish mind" all combined to divert attention from the simple arts of husbandry that might have insured a sounder, if less spectacular, future for Spain than has been her lot.

North of the Mediterranean the land likewise fared ill. Neolithic farmers had practiced a system of clearing, burning, farming, and abandonment that works well enough until population becomes so heavy that the time allowed for recovery and repeating the cycle becomes too short. The breakup of the Roman Empire gave way to feudalism with its contempt for manual labor and learning, along with its heavy emphasis on warfare.

Here the story of land management is confused by the presence of many small states, each with a complicated course of changes in the system of land tenure and farm labor. Nearly everywhere, however, with increasing wealth and organization, the holdings became larger and the status of the actual worker worse. The manufacture of steel required fuel in huge amounts, and the forests were early cleared to provide wood for the charcoal burners. This took place before the land was needed for agriculture, and long before the demand for lumber itself would have caused the cutting. The use of coal and coke in metallurgy is, of course, a very recent practice.

By the time of Charlemagne, who was an enlightened ruler, the onslaught against the forests of western Europe was under way, to continue through the thirteenth century. By the end of the Middle Ages the land was largely divested of its trees, as the Mediterranean region had been before the Christian era, and stringent laws against cutting came into being. Whatever advanced ideas had been inherited from Rome were soon lost to sight. Fields were used, then abandoned. Feudal lords shifted their headquarters from one castle to another, to get away, it has been said, from the accumulated filth. But the coefficient of toleration for filth was so high in those days that the moving was more likely to have been for the purpose of tapping new sources of food as the old sections of the fief played out. Eventually, after a period of rest, the abandoned fields had to be used again.

Such a system is unsound. Recuperation takes too long, and too much of the land at a time remains idle. Paintings and sculptured figures of the period portray human beings who are wan and rickety, and since these portrayals were most common in sacred art, we are still inclined to feel that anemia and sainthood are inseparable. Actually the trouble was caused by inadequate diet and malnutrition on a huge scale, such as we sometimes find in backward rural communities.

In contrast to the great landed estates, church property seems to have been well managed. As to the monks, even their friends are obliged to admit that they were well fed. In modern England the sites of old monastery gardens are still marked by little patches of savory plants, and such curiosities as the fat-leaved houseleek, or hen-and-chickens, which the good monks used to grow and pickle for winter use. To what extent the landwise thrift of the church was due to its discipline and abundant resources, and to what extent to a knowledge of Roman agricultural lore, it would be interesting to know. But to revile the monasteries for the benefit they received by keeping alive the tradition of good farming is as unjust as to censure the knowledgeable farmer of modern days for prospering when his neighbors fail. In either case the slow effect of example is the surest road toward the common good.

The stripping of the western European forests not only produced a shortage of timber and fuel, but also caused other disturbing consequences, such as puddling of the soil and formation of acid heaths, particularly near the Atlantic. The result of this was to force upon Europe the early development of public and private forest policies—a scheme that incidentally fitted the idea of great royal and manorial hunting preserves.

Meanwhile the dawn of better husbandry began in the Netherlands, where land reclamation was under way in the thirteenth century. With it developed a return to legume rotation, the use of manures and other fertilizers, and the growing of root crops. The counsel of Virgil in his Georgics to "sow golden spelt, where before thou had reaped the pea with wealth of rattling pods, or the tiny vetch crop, or the brittle stalks and rustling underwood of the bitter lupin" had long been forgotten in a world of equally illiterate nobles and serfs.

Scant yields of grain (10 bushels of wheat to the acre) and forage under heavy pressure made the overwintering of livestock difficult. Hence the feasts at winter solstice and the scanty rations of late winter converting grim necessity into the Lenten virtue of fasting time. Root crops, beets, turnips, and their allies, brought in by the new lowlands thrift helped mightily in solving the problems of animal husbandry.

During the fourteenth and fifteenth centuries these reforms spread throughout the continent of Europe. But sad to say, they had not reached Britain or Spain by the time the Americas were colonized. They also failed to get beyond feudal barriers in East Prussia. We do know that space in the Netherlands was at a premium, people were many, and the economy was flourishing. Food *had* to be available, and enterprise was in the air. Shadowed in the background of challenge and opportunity are two questions, at least for the pedant: to what extent had Roman experience been recovered and, if it had, whether it had been made available through monastic example or the revival of classic learning.

At any rate it is an old axiom of agricultural history that the greatest progress comes where man has to be diligent in order to survive. The intensive skill required to reclaim and handle the lowlands proved to be as great a stimulus as irrigation had been in other parts of the world. The result was that with the renaissance of the spirit throughout the western world, there was also a renaissance of sound land management in this area. Not only were legumes regularly included, but rotation systems that employed root crops were worked out, with immense benefit to man, beast, and land. These practices were not substantially improved until the day of modern, scientific, intensive rotation.

The awakening interest in science—"natural philosophy"—helped spread these ideas and caused a search to be made for laws and explanations. The Dutch and Flemish taught their neighbors how to farm. This happy service came at a time when expanding populations taxed the existing food supply. Interest was further stimulated by the introduction of curious and valuable plants from the New World. Sometimes these plants reached western Europe in a roundabout fashion. The French still call maize or Indian corn by the name *blé de Turquie*, or Turkish wheat, and of course the domestic turkey itself is an American bird. The word "potato," now applied to a South American plant, can be traced to a confusion with the West Indian *batata,* our yam, while the prefix "Irish" has nothing to do with the real origin of the plant.

Incidentally, the potato is an important illustration of the role that fashion played in the revival of agricultural practice in Europe. Statesmen realized the importance of this plant as a source of cheap and needed food for the poorer classes, but were obliged to use great ingenuity to break the chains of prejudice against its use. The Grand Monarque wore a sprig of potato blossom at his lapel, and served the tubers at a stately banquet. Another enterprising gentleman guarded his patch against theft with ostentatious care until the crop was ready to dig. He then relaxed the guard and human nature took the next step, permanently inoculating the surrounding peasantry with a taste for this habit-forming vegetable.

Under Charles II the seal of fashion's approval was set upon scientific experiment by the founding of the Royal Society of London. Slowly an understanding of the relationship between soil, water, sun, air, and plant life developed. Modern chemistry, which had its beginnings about the time of the American Revolution, helped the cause along immeasurable. In 1838 the testy German chemist Baron von Liebig began his search for the perfect fertilizer, one that would be applied artificially and forever do away with the problem of exhausted soils. Like the alchemist's search for the philosopher's stone and the elixir of life, his dream was never to be realized in an absolute sense, but from his efforts came a better knowledge of both plant and soil. The great agricultural experiment station at Rothamstead, in England, was founded about 1843 as a private enterprise by Sir John Lawes, with the help of the scientist Gilbert, and its fields have been under continuous experimental control ever since.

Louis Pasteur's investigation of bacteria and other forms of microscopic life opened up new vistas of knowledge regarding the soil and its processes. Modern physics was drawn upon to explain many things that had been mysterious about the relation of water and soil. The ancient custom of legume rotation was proved to be a means by which nitrogen from the air was fixed in the soil through the work of bacteria growing on the roots of clover and beans and the practice justified as the cheapest, readiest means of supplying this chemical substance. In dark, immense old Russia, whose little handful of scientists has at times seemed gifted with an eerie second sight, great things were done. The basic principles of soil development were worked out there. We now know that soil is not a substance, or a mixture of useful chemicals, but a phenomenon of the utmost complexity, whose delicate balance is easily disturbed and whose complete interpretation is yet far off.

The meaning of all these discoveries in a few words is this: The inexorable laws of cause and effect operate in the production of food from the soil just as in every other realm of physical experience. No man, no nation, can spend resources faster than they are built and escape the inevitable reckoning. It is impossible here, as elsewhere, to get something for nothing, and supreme folly to trust to the future for our errors to right themselves.

Modern Europe, still struggling to reconcile the advantages and social cost of the Industrial Revolution, with political ideologies serving more to rationalize the struggle for power than to promote the humanitarian values they profess, exhibits a patchwork of good and bad husbandry of the earth. Fossil energy, harnessed to mechanisms that can endlessly manufacture consumer goods out of raw materials, has seemed to promise an end to the ancient specter of want. Improvement in production, transportation, medicine, and other products of applied science have produced an increase in population unmatched in history.

France, less from new knowledge than from laws that encourage the keeping of real estate in the family, had reduced its birth rate. Much later Britain, aided by the publication in 1922 of a book on contraception, was attempting to stabilize population growth. Scandinavia, some of whose adults spend much of their lives at sea, along with other small areas of the continent, has made admirable attempts to adjust numbers and practices to the limitations of space and resources. Much of the success of Britain in surviving World War II can be credited to a survey of land-use capability initiated by the geographer Sir Dudley Stamp and carried out by school students after World War I.

Along with other "advanced" or "developed" cultures, those of Europe have become committed to the values of mass production, despite its pitfalls when applied to living organisms, to say nothing of education, where its sad and costly failure is now apparent. In East Anglia mechanized farming without livestock, after an initial success, has shown a sudden drop in yield. This has been traced to a loss in the crumb structure of soils, a condition necessary to air, water, and nutrient controls. Under normal conditions crumb structure is maintained by the presence of organic glues (colloids) provided by animal wastes. Instead of a return to a mixed plant-animal regime, search for a chemical that would restore crumb structure has been proposed!

Again, the late Nikita Kruschev, whose good will and desire for a solvent world peace has probably been underestimated, was deeply impressed by his visit to our corn belt with its heavy yields. To what extent he registered the cost in terms of heavy machinery, fertilizer, fuel, and other petrochemical products, we can only guess. We do not know whether he was reminded that in some cases more energy goes into an acre than is returned in crop yield and that less than a thousand miles to the west the wholesale use of mechanized, mass production of wheat had led to failure and vast clouds of dust in the 1930's.

But we do know that he repeated our mistakes in the semiarid East of his own vast country and that the resulting failure was a major element in his being converted to a nonperson under a political system that gives no blue ribbons for effort.

4

Poor Richard, Poor Lo

"THE ONLY GOOD INDIAN is a dead Indian." It did not take the children of the Pilgrim fathers long to forget that the colony that landed at Plymouth Rock had been kept from starvation by friendly Indian help and tutelage in the art of New World agriculture. Old Squanto as an individual may be accorded the honor of perennial mention in history books, but the cultural and industrial achievements of his people were methodically omitted, or else cruelly distorted. Thus the same generations that were nourished on the *Book of Martyrs* and steeped in pity for white men who died for what they loved found no difficulties, except those of a technical character, in crushing the red man and his works. And when, about seven generations after Plymouth Rock, all Indians were evicted from the beautiful rolling lands of northern Ohio, the substantial citizen who ambushed a lonely stray and shot him in cold blood received little more censure than if he had killed a coyote. Between the English and the Indian it was war without quarter almost from the start, with the end inevitable.

The reason is not far to seek. With that same granitic serenity wherewith the Victorian transplanted his tub, his cricket, and his teapot, the first Yankees were bent upon moving their British world along into the New World with them. And in particular they brought their women, joined to them in lawful wedlock, thus introducing the fateful factor of competitive breeding. Not only did this intensify all the effects of invasions and conquest, but it provided a certain sanction for every aggressive move against the aborigines. One of the most significant traits of Anglo-Saxon psychology is the need for lofty motives during the process of getting whatever may be wanted. More a capacity for self-deception than a matter of perfidy or hypocrisy, it nevertheless lies at the root of much of the mischief done to the resources of the North American continent.

True it is, the clear-headed Latins came west with their own sanctions, extended by Holy Church. And while these did not suffice to prevent cruelty and greed, there was this beautiful element of consistency: A race that was good enough to furnish converts was good enough for intermarriage. Doubtless this process of amalgamation was the easier because so many of the Romance invaders came as soldiers and traders. At any rate, where they settled, the Indian and a great deal of his culture have survived.

The Englishman regarded with disfavor the Indian custom of allotting the agricultural labor to the woman, and pictured her as a slave. The fact is that she was not only at least as good a farmer as the average early English settler, but was happy in her work. The brave for his part had the responsibilities of war and chase. The fact that he was exterminated rather than enslaved carries its own comment on his character. It would be an interesting digression to compare the lot of the Indian woman with that of the white wife who was brought along on the first dark ventures into unknown perils and hazards. Yet it must be conceded that if the latter suffered, her daughters have come into their own. They now have title to a generous majority of assets in the United States, as well as preponderant influence in many cultural activities that their men have been too preoccupied to attend seriously.

The Indian culture that the Englishman destroyed along with its creators was diverse. In some respects, it was on a level with that of his own Stone Age ancestors, in others far above. Of necessity it was in balance with nature, since the Indian had to adjust himself to nature rather than compel it. The Indian had a very keen sense of his dependence upon nature, and an aversion to needless waste of its resources, a fact that some of his sentimental friends like to dwell upon. But with his technical limitations it could not be otherwise. The mound builders, whose elaborate and interesting works are found in Iowa, Wisconsin, and Ohio, represent a comparatively high type of culture that at one time spread out to the limits named from some southwestern center. However, they could not maintain existence in the face of difficulties, probably caused by the encroaching forest with its marauding groups of hunters. As a result their way of life had vanished before the whites arrived. Their technical control of the environment was too slight to succeed permanently in peaceful farming villages, and they disappeared. Given steel tools for work and war, domestic animals for labor and food, the tale might have ended differently. It is well to remember that the Indian had cultivated all of the New World plants that have been found to be worth the trouble and had used excellent methods in doing so.

A feature of the Indian's close relationship to nature was his lack of any sense of private ownership of the land. The land was there to be used, and the general right to its use had to be determined, on occasion, by bloodshed. But its use was for the subsistence of the group, not for private gain. In this sense, it has been well said that in the Indian land agreements with the white, the red man sold one thing, the white bought another very different.

To the Englishman, on the other hand, the right of a commoner to private ownership of real estate was a serious matter, the fruit of centuries of struggle against a powerful social and economic system. Not only was his house his castle, but his lands were his to use, enjoy, exploit, or ruin, as he would. The only checks were the requirement that he pay taxes levied by his own lawful representatives, and the provision of escheat, or return of the land to the government in case its owner died with no heirs.

The first settlers in New England were largely townspeople, not skilled in the art of farming. Even had they been mostly country born, it probably would have made little difference. British agriculture of the day had not yet learned its lessons from the efficient lowlanders. It involved problems of soils, crops, and climate more or less different from those encountered

by the settlers, and was organized on a manorial basis
for which there was no use in New England. On the
other hand, the settlers were ingenious craftsmen, industrious, thrifty, and determined. Had they been
without these qualities, they could not have survived.
They spread out over the land, it is true, and made
their own subsistence from the thin, rocky soil. But
their real energy went into manufacturing, marine enterprise, and business, ending of course in the severe,
direct, commercial rivalry with the mother country
which brought on the Revolution.

South of New England in New York, Pennsylvania,
Maryland, and Delaware, not only was there a higher
proportion of good farming land, but many more
settlers from the continent of Europe. Here the late
medieval reforms in land use and management had
spread from the Netherlands by the time North
America was colonized. Thus a sense of good agricultural practice was ingrained in German, Dutch, and
Scandinavian newcomers. Their practices soon became, and still remain, models of their kind. Their
farms were beautifully kept. Fortunately, the growing
cities of the seacoast afforded a ready market, permitting money to be used in maintaining fertility. If conditions everywhere were like those of the fruit, truck,
and dairy regions of eastern Pennsylvania, Maryland,
Delaware, and central New York, there would be no
deserts on the march today.

Scarcely had the Revolutionary War ended when
surveyors, and after them settlers, began moving into
the country north and west of the upper Ohio River. By
1800 the movement was well under way. All elements
of the colonial population, save perhaps the great

southern manorial owners, took part, although even the latter were involved as capitalists. Increasing numbers of emigrants from Europe swelled the tide. The only challenge apparent was that of land occupancy. Water-mill sites were prized locations.

The eastern third of North America was covered with magnificent forests, the result of an excess of rainfall over evaporation. Here and there were grassy glades that became larger and more frequent from Ohio westward until they expanded into the vast domain of the prairies. Convinced that the absence of trees was a sign of poor soil, the first settlers chose to clear the forest for farms, using the grassland with its richer soil for pasture.

The forest itself, although a welcome source of fuel and timber, as well as game, was regarded principally as an obstacle to agriculture. Not only was wood used lavishly, but the felled logs of valuable, even precious wood, were rolled into huge piles and burned. Walnut and wild cherry fed the flames along with elm and cottonwood. Perhaps this was inevitable, but it is worth noting that the field books of the first surveyors contain abundant evidence that they knew the value of the forests through which they worked. So far as anyone can determine, these records were never utilized in any respect for public good save as that end might be served through the foresight of the individual settler or the speculator who followed in his wake. When we recall that the center of the commercial lumbering activities did not move west of New York until after 1850, it is easier to understand the apathy of the early nineteenth century toward forest conservation. Not only great sagacity, but a greater immunity to political pressure than our statesmen commonly show would have been required to establish the reserves or restrictions that we now know should have been made for the common good. And it may be doubted whether such precautions would have been worth the trouble, in view of the squatters and other lawless gentry that abounded. It is no secret that the Whiskey Rebellion that began in Washington's administration continued to a climax during Prohibition and, along with other illicit activities, is not yet ended.

Other circumstances were unfavorable to what we might call a psychology of permanence. In the beginning the urgent need was to meet each day's practical problems as they arose without worrying much about the future. Later the processes of urbanization and industrial change occurred so rapidly as to be bewildering. It took a strong character, or an utter lack of enterprise, to avoid riding the tide for profit's sake. To a degree even the best citizens conducted themselves like the shell-game man at the county fair, on the basis of quick action. Yet it is only fair to note that many monuments to the future in the way of schools and other humane institutions had their origin in this period.

The resourcefulness of the early rural population was remarkable. The farms of that day were self-supporting to an extent undreamed of now. Home-grown wool and flax were spun and woven at home, and there made into clothing. Light came from homemade candles, cast from beef and mutton tallow. These had their faults, as anyone who has inhaled the aroma from incandescent mutton fat will know. But they cost no cash outlay. The tiny driblets of cash, which a modern family would despise, were hoarded, often to be used in the purchase of additional land. Many a family that had struggled through years of toil and painful saving found itself at last wealthy, for the value of land steadily increased as population developed. Good business management was held in high social esteem, but there was no corresponding pressure to encourage wise husbandry of the land. Land was land, to be mined until it could be sold for a profit.

The professional classes were on the whole an excellent group. Generally prosperous and often public spirited, they were at once an influence for stability and unrest. The example they afforded did much to drain the farms of their ablest young men. The unremitting toil and lack of conveniences, let alone luxuries, that the farm implied were constant incentives to escape. Added to this must be the temptation to retire to town or village that assailed farmers and their families who had continued the struggle long enough to become wealthy. That this kind of translocation did not always succeed is witnessed by the old proverb "one generation between shirt sleeves and shirt sleeves."

Despite the rapid industrial progress, there set in a serious decay of rural life. The land, already exploited rather than conserved, was turned over to tenants, generally without the protection of just and intelligent contract, and in half a century the vicious cycle that, spreading over centuries, had ruined Greece and Rome was on its way. The tenants, like their owners, lacked the steadying effect of a long agricultural tradition. Such apprenticeship as they had was in what was, at best, a predatory kind of farming. Thus it happened that while people in the Middle West were discussing the tragedy of abandoned farms in New England, their own land was moving rapidly toward the same fate.

We have seen that the center of lumbering lay in western New York as late as 1850. Thence it shifted to Ohio and Indiana, and then quickly into Michigan, Wisconsin, Minnesota, and since has moved to the Gulf and Pacific coasts. In every instance until recently the method was that of clean cutting, with no thought of permanent yield. This policy was applied with equal vigor on the land that had farming possibilities and on land that had none. Rather generally the best potential farmland was occupied by farmers before clearing, and the lumber sold from it by them. The great holdings of land suitable only for forest were stripped by the companies in control, after which attempts were made to colonize farmers on them, often with sorry consequences.

Following on the heels of the main lumbering activity were numerous industries that utilized second-growth timber and remnants. Stave and hoop mills, bending works, wagon and handle factories may be mentioned among these transient industries that shifted west in a slow afterwave. They represent merely one of the many proofs of instability. The psychology was not one of a settled order. People never felt that they must face their immediate surroundings with any finality, even though they might have passed most of their days in one spot. If things became too bad, moving was always possible, and land elsewhere was still cheap. In consequence the obligation to conserve the soil was scarcely felt, and the farm itself came, as we have seen, to be regarded as a mere stepping stone into the professional or commercial life of the rapidly growing towns.

Meanwhile the developments of commerce, fuel, ore, and manufacture, and the sequence of wars, booms, and panics in the Middle West served to obscure the serious character of the agricultural tragedy that was being enacted. "Our Uncle Sam is rich enough to give us each a farm," ran the old song. The retired Ohio farmer could take his profits and put them into cheap land in Iowa. True, there were no railroads in northwestern Iowa, but after the land was acquired, a word with an old friend in the United States Senate—all perfectly honorable—and the branch road sprouted out in the proper direction, to the good of everybody concerned.

Probably Iowa deserves as much credit as any state for arresting the blind progress of pillage. There are older experiment stations and agricultural colleges than hers, possibly even better ones. But nowhere before had such institutions enjoyed such public confidence and been so concrete a factor in the state's policy as in Iowa. Here at last was a state in which agriculture was not merely one ring, but the whole circus. The result was that Iowa land was well-managed and commanded a continuously high price, even after much of it, too, was actually handled by tenant farmers. Beyond question this was largely traceable to the influx of Scandinavian, Dutch, Bohemian, and other European continentals experienced in the skillful and careful husbandry of their perfectly managed European farmsteads.

Farming in Iowa, and in the parts of Kansas, Nebraska, Minnesota, and the Dakotas known as the grain belt, enjoyed a definite status. Instead of a makeshift or an outcast occupation, it was regarded as a prospering, well-managed enterprise. This position was not won without a period of false starts and discouragements that eliminated all but the most successful and persistent, especially west of the Missouri River. In Wisconsin the portions suitable for dairying became specialized for that industry and still occupy a premier position. In the cut-over forest land of the northern lake states, however, attempts, often strongly sponsored by the exploiters, to convert such land into farms were largely unsuccessful. In the end, they proved very costly, not only to the individual colonizers who failed there, but to the eventual profitable reforestation of that region. It is worthy of note that the ultimate failure of these schemes was predicted by competent agriculturists. At least one such man was forbidden to make public statements about the matter.

It would be an error to say that the soil of the grain belt has maintained its fertility undiminished. Despite better-than-average management, much of its rich three to five feet of grassland soil has been washed away. In the spring of 1962 the waters of the Cannon River flowing south through Northfield, Minnesota, ran inky black with its load of prairie topsoil—by no means a unique instance.

Timber claims were started in many places, somewhat tempering the individual homesteads and fields against wind and sun. The worst error of management has been the failure to keep an adequate reserve of

land unplowed as pasture or hay meadow. Most farmers of the grain belt would today be better off than they
are with perhaps a fourth of their land in productive
pasture—to provide them with their own milk and
beef, if nothing more, while hedging the risk to soil and
water supply on fields that are tilled. Ominous for future soil condition is the widespread absence of livestock on farms now highly mechanized for the production of cash crops.

Furthermore, the pattern of farm finance that developed west of the Mississippi was thoroughly vicious
and against the public interest. This is not to say that
any individual responsibility can be fixed. By the policy
the banks profited hugely for many years at the expense of their own clients, but in the end the banks
crashed with the rest of the system. The wrong of this
system was the failure to appraise land on the basis of
its average yield. This is particularly serious matter in
an area where the climatic margin of safety for good
crops is never large, and where the rainfall fluctuates
greatly. There are certain to be years of good crops and
years of failure. In good years everyone who has some
money saved wants to buy a farm. Under such conditions the banks permitted their customers to pay more
for land than, on a sound average, it was worth. Inevitably came the years of crop failure. Payments
could not be maintained, and the bank that had advanced the money—after receiving a large down
payment—got back the land and, of course, retained
the payment. It was not uncommon for a bank to resell
the same piece of land four or five times, each time
with a handsome profit at the expense of some client's

life savings. The fisherman who has to use a fresh worm for each catch will appreciate the remarkable features of this system, complicated as it is by continuing inflation and the rapidly decreasing per capita space.

Of course, land values ought to be computed on an actuarial basis, as life insurance premiums are, even though this task is the harder. But it would be a distinct improvement if banking ethics did not permit a customer to pay more for land than the banker himself would care to pay in order to receive adequate returns.

The northern farmer, then, moved across the continent with the complete destruction of native Indian culture, emphasizing at every step individual property rights in the land without regard for public policy. Until the grain-belt states were reached and self-interest dictated better measures, the farming areas were systematically impoverished and the lands exhausted.

Within the grain belt, although a high level of soil management was inaugurated, too small a proportion of native grassland was retained as a reserve and safeguard against climatic fluctuations and soil removal. Before 1930 a system of farm finance that was essentially unsound had developed, impoverishing the farm population and ultimately the financial institutions that served the area. Now, in the final quarter of the twentieth century, we are reminded that population increase and inflation leave us with only wealth or inheritance as means of acquiring good farmland. Soon only the former will be effective because of almost confiscatory estate taxes.

We have followed the European west through the

northern states and into the eastern portion of the great interior grasslands. Before moving farther west into the shortgrass plains, where much of our story lies, let us retrace our steps to the Atlantic coast and work westward once more, this time with a different breed of men, the southern colonists.

5

The Southland

LONG BEFORE THE WHITES moved west from the Atlantic coast, their pressure on the Indians had caused serious disturbances far inland. In particular, the rich, though rough and broken, region that is now Kentucky and West Virginia had been depopulated and was serving as a no-man's-land separating hostile tribes with distinctive types of culture. The southern tribes, increasingly toward the Gulf, displayed a settled village economy with highly developed agriculture. They possessed extensive earthworks, storehouses, various crafts, and commerce. Plainly they showed signs of influence radiating not only from Mexico, but from the Antilles, and because of the genial climate, they possessed a rich variety of cultivated plants.

Curiously, the same dark and bloody ground that had separated the Indians was destined to cleave the white Europeans into two profoundly different cultures. North of it was the strongly individualistic middle class from the Old World, each member conscious of his own hard-won rights and the new freedom to exercise them in a region of unbounded resources. In these people considerable practical and mechanical sense was combined with the habit of hard work. Except for the Teutonic groups among them, they were not, as a rule, first-class farmers. We have seen the result in a trail of impoverished land extending west to the grain belt, but accompanying it a brilliant record of commercial and industrial growth.

South of the dark and bloody ground, the picture grows more complex. Here the gentlefolk of old England were granted huge estates. With them as workers came the impoverished, the inefficient, and the failures from various levels of society. Black slaves, soon to be proved unprofitable in the northern states, were found to be decidedly the reverse in this warmer region. Tobacco, cotton, indigo, and sugar cane could be grown in addition to staple food crops, and were as readily convertible into cash as the products of northern industries. And like the factory products, it was soon discovered that they could be most profitably and economically produced on an extensive scale. Just as in the North the small independent shop was to give way to the large, centrally managed factory, so in the South the small, personally owned and worked farm could not compete with the large, slave-operated plantation.

Along with this circumstance was another, one that was destined to cost a staggering total in the ensuing centuries. The wealthy controlling group classed education along with the other good things of life—a luxury to be dispensed only to those who had the means to pay for it. There is no use at this late date to chide them. This position seemed as logical and just to them as some northern judgments seemed outrageous. The poorer class of whites, driven out of owning the richest lands by competition, and kept from employment on these lands by black labor, were also kept in ignorance. Furthermore, since the mechanical duties in connection with the larger plantations were cleverly performed by trained blacks, there was little opportunity for the whites to become, through apprenticeship, skillful artisans like those of the North. In the end, the blight of ignorance rested not only upon the impoverished white misfits of the lower classes, but even upon the poorer kinsmen of the wealthy landowners.

The best land in the South lay along the coastal plain and in the rich, broad valleys that ran inland from it. This, if course, came under the control of the great landlords. To survive, the poorer whites had literally to take to the hills. And while the seacoast, both along the Atlantic and the Gulf, became a region of wealth and culture, the interior developed its own discipline and its own significant character. The line of cleavage within the South was as distinct as the cleavage between North and South. Had the Civil War not intervened when it did, the South eventually might have witnessed a civil war of its own. For the people of the interior hills came to realize they had one weapon, the ballot, and by 1840 they were using it with telling effect. In Mississippi and Louisiana they forced the removal of the political capitals away from the coast, and into the interior.

Meanwhile, the Indians had been largely cleared out. Some few in the Carolinas managed to hang on, adopting the culture of the poor whites. Seminole bands found refuge in the Everglades. Farther west, among the civilized tribes, the plantation system, with slaves and large holdings, was put into operation. But throughout the South, which, with the exception of such factory regions as Birmingham and the great shipping centers, was predominantly agricultural, the custody of the land fell into the hands of the white race. How was this trust discharged?

Despite their faults, the landed gentry had great virtues. Bright among these was the strong sense of public duty, even though it might not always extend to popular education. Under these circumstances good land management and enlightened agriculture became a moral obligation as well as being good business. George Washington exemplifies this attribute, and often hidden behind his better known titles is that of "Father of American Agriculture." He was a notable student of the subject, diligent in the handling of his land and concerned with introducing the best possible methods. Despite his example and efforts, which were repeated through all of the plantation country, the fertility of the coastal plain and great valleys was steadily depleted, so that today their fertility can be maintained only by costly applications of fertilizer.

Perhaps this would have been unavoidable in any event. The soil layer is generally light, underlaid largely by sand. The constant demand for cash crops made it extremely profitable to grow the same crop repeatedly, even if the expense of fertilizer was necessary. Anyone who has watched the ash grow at the end of a burning cigar may realize how much material the tobacco plant takes from the soil. All of the ash, and not a little of the smoke, represents stuff drawn in by the roots.

But the main trouble with agriculture in Washington's time was the same thing that handicapped the medical practice that bled him to death in his last illness. Both were looking to science for light before science had the light to dispense. Under such conditions the most conservative position is the safest. In medicine the physicians of the Middle Ages had developed many practical measures they could not explain, but that had the merit of working. For example, they burned fresh wounds with a hot iron. At the time of the French Revolution all of these old practices were cast aside as being superstitious. Medicine attempted to be scientific with awful results. Instead of cauterizing fresh wounds, doctors bandaged them in lint, which we know must have been bacteriologically filthy. Gangrene was the usual consequence. In Washington's day the relation of soil, air, water, and sun to plant life was just beginning to be understood. Some of the aspects most important for agriculture were not discovered until a century later; and of course there is still much to learn.

Had the coastal plain been farmed throughout by industrious peasants, drilled in the routine of the best agricultural tradition of Germany or Flanders, it is probable that it would have remained more productive than it did under its relatively ambitious management. The peasants would not have been able to explain the scientific basis for their practice, but the practice itself would have been sound.

There is today in South Carolina an area that has long been occupied by Germans. All around it is desolation caused by erosion. This erosion, of course, was induced by improper soil management. Within the area in question erosion is comparatively slight, although the physical character of the land is like that of the surrounding region. Nor is this cultural enclave of sound traditional practice unique; among others has been the Norwegian settlement at Cranfills Gap, upstream from Waco in Texas.

Not all of the coastal plain was given over, of course, to plantations. Toward the south the extensive areas of longleaf pine yielded timber and naval stores, while the swamps contained valuable cypress and gum. It is almost unnecessary to remark that the development of these forest resources was, until very recently, a matter of exploitation rather than conservation. The situation was aggravated by fire, which in some instances apparently produced barren areas largely unfit for any use. The turpentine trees were tapped by boxing, which quickly destroyed their value, either for turpentine or timber, and made them an easy prey to the frequent burning. Properly used, however, fire is important in forest management.

Above the coastal plain lay the foothills and mountains. In some areas there were broad and fertile valleys, but more often the fertile valley land lay in narrow isolated strips. Hither the poorer white population was forced to retreat in order to survive. Hill and valley alike were wooded, with a rich variety of trees. To the settler, here as in the North, the forest was a hostile thing, occupying the ground that he needed for corn and beans, even though it furnished him with game, fuel, and building material. All was fair in the struggle against this handicap, and no weapon, not even his sharp ax, was more powerful than fire. So the use of fire against the forest became a part of the ritual of the poor white. He has literally burned his way west, from the pinelands of the Carolinas to the blackjack cross timbers of Oklahoma and Texas.

The economy thereby introduced was one of clearing the forest by fire, farming the cleared place as long as it yielded well, repeating the process until it became necessary to move to a new location. This is Stone Age technique, of course. Practiced by the ancient Mayas, it defeated them. Practiced by the hill people of northern India, it keeps them on the move, and keeps down their population. And in western Europe it was long ago abandoned in favor of more permanent methods.

No less than the wealthy planter, the mountain white possessed his virtues along with his vices. A common error is to regard him as being of inferior metal biologically. Of course, as may happen in any mountain region, for example Switzerland, there have been isolated districts in which inbreeding, combined with ignorance and disease, has produced genuine physical deterioration. But for many years it has been known that no finer soldiers are found in the United States army than those recruited from the hills. Given proper influences, the young people develop into excellent, even distinguished citizens. Witness the thrifty young products of modern 4-H clubs, not to mention Lincoln and his successor. Anyone who has traveled among the mountain whites is familiar with their fine, instinctive courtesy and general decency. The fact that their women lead lives of hard work and wear out quickly is often misunderstood. Their women are held in high regard and the division of labor is a necessary one. These people are not fundamentally lawless. It is true that laws made and administered out of their sight have little force. But their own code is definite and strictly enforced, even though the means used may be extralegal.

A young engineer from the North who has lived among them for a number of years says:

When I first saw how hard they worked and how little they had, I thought they were to be pitied. Now that I know them better I think they lead much happier lives than the rest of us. They are not lazy. Those whom you see sitting in front of their houses at mid-day were up before dawn and had done a good day's work. They are not dirty. I have eaten with them frequently and often shared a one-room cabin with a family of six or seven at night. The children, even the older girls, undress without self-consciousness down to clean undergarments and retire, as do the boys and men. The mother usually waits to put out the light, then prepares for bed. They are fine, decent people.

When the depression of 1930 struck, it caught them at first unprepared, as it did everyone else. But immediately they took steps to store up reserves from their excellent gardens and the abundant wild fruits about them, as well as their casually managed livestock. The second winter of the depression found them in tidy shape. True, there was government aid—and even a rumor or two that relief sugar, ostensibly for canning, was being converted into a more marketable product—but on the whole the mountain whites weathered the storm. The very simplicity and self-contained character of their living made it possible to do this with relatively little adjustment. An anthropologist who visited them for the first time said: "I have found a neolithic culture. The metal tools and manufactured junk they have are only incidental, and merely emphasize the self-sufficient, primitive economy under which they live." Without any great taste for mechanical devices, they nevertheless possess an uncanny skill with simple tools. A cabin in the Cherry River district of West Virginia was built, fenced, neatly and conveniently furnished, and all the work was done with saw and ax—mainly the latter.

So long as these lands are not too densely populated, and the uplands around the valleys are richly wooded, the people can maintain themselves. But if lumber companies clear the woods, or the farmers persist in burning them to improve pasture, or worst of all attempt to farm the hills, the end comes quickly. The hills erode and the small, rich valley fields are buried under the waste. The skeletons of deserted cabins are still not uncommon in the southern hills.

From the most hopeless of the mountain whites, and the doubtless other sources, comes the derelict white humanity of the South. Mill hands, itinerant cotton pickers, and forlorn tenant farmers are included here. They are in direct economic competition with the blacks. If not on the move, they are found as workers on neglected land, held by absentee speculators for one reason or another. Their plight has attracted enough attention in high places to spare its discussion here. They are not to be blamed if the land they occupy is skinned and ruined. Few are trained as farmers. Few have sufficient capital or income to manage well if they could. The owners of the land on which they live are more often than not purely predacious, and the land generally is run down when they move onto it. Their pastures are loaded to the limit. The timber is stripped off and hauled in to sell for fuel. Any land that promises even one or two good crops is likely to be cleared and plowed, regardless of the aftereffects. They have few milk cows, and those inferior. There is a shortage of manure for the fields, and no money for commercial fertilizer. Frequently, vitality is so lowered by disease, malnutrition, and bad heredity that it is hopeless to expect the energy that would be needed for good management even if all other essentials were provided. In these human derelicts we see the cumulative effects of three centuries without popular education, social discipline, economic opportunity, or good craftsmanship.

As tenants, they have been crowded off the well-managed farms of the South in favor of dependable blacks or enterprising whites. They are to be found on marginal and submarginal land, most of which in the East should be back in forest, in the West in grass. The zones they occupy can be detected from the air by the red or yellow gleam of sterile, eroded hills. Although their hands may have hastened the destructive processes of disturbed nature, the responsibility to society is one that, in all justice, should not be laid to their charge.

Through the pine forests of the Gulf states and the oak forests of the Ozarks the white man debouched upon the southern part of the great interior grassland. Here, south of the grain belt, is a rolling terrain of varied soils and possibilities. It was leased in great holdings of thousands of acres by the cattlemen, and the coming of homesteaders was viewed by at least one of the latter with complacency. "They cannot farm this country and live," said he. But farm it they did. The valleys were rich and fertile. Even the uplands gave good crops at first—wheat, oats, corn, sorghum, and cotton.

The fire god was invoked to clear what woods there were and to improve the shrinking pastures. In the hands of the old cattlemen fire had been used, if at all, with care and skill. Now it was brandished recklessly, destroying humus and killing valuable kinds of pasture plants. Pastures were overloaded. Upland fields were carelessly planted in row crops. In a few decades, save for the exceptional, well-managed holdings, the once-green countryside showed everywhere the scars of soil destroyed, and even the rich valleys became nonproductive.

Once again we have reached the eastern portion of the great grassland, this time by the southern route. We have traveled from the fair valleys of Virginia, which even the foresight and care of Washington could not preserve from loss of fertility, to a region into which man has swept like a devouring plague, reducing much of it from a prosperous cattle country to a region of deserted and deteriorating farms—in four decades.

In the more than two-score years since the preceding paragraphs were written, vast changes have taken place in the southeastern quadrant (about 24 percent) of the contiguous forty-eight states. Many were long overdue, delayed by a savage refusal to adopt Lincoln's wish to extend a welcome to the seceded states, which instead were treated as conquered provinces. Also involved was the cultural inertia of a social and economic system that had been crystallizing for two centuries. Northern critics seldom realize what a different course their own zeal might have taken, given a different birthplace.

The plantation system of the South, worked by slave labor, was curiously parallel to the kind of land use that led to disaster in Greece and Rome. Conditions of soil and topography in the southern Atlantic coastal region sped up the effect of cultural practices in hastening the formation of great gashes called gullies. Even in colonial times these evidences of erosion caused by misuse could no longer be overlooked. The first recorded concern over soil erosion came from the South. Meanwhile, the usually canny New Englanders insisted that the glacial boulders newly exposed each year as the tilled soil above them washed downhill had pushed up during the cold of winter! There was just enough truth in this self-deception to convince even some scientists until recently.

The South furnished not only the first and most dramatic evidence of soil erosion, but also the man who did most to bring it into focus. Called Big Hugh and known for his size, appetite, and gusto, Hugh Hammond Bennett (1881–1960) was reared on a North Carolina farm and, after graduation from the University of North Carolina, became a soil surveyor in the United States Department of Agriculture. At the time, soil erosion was either ignored or dismissed in official circles as of little significance. In 1909, Bennett's chief pronounced soil to be the one inexhaustible resource, causing Bennett later to remark that he did not know so much misinformation could be compressed into one brief sentence. And in the same decade a warning to

the secretary of agriculture that overgrazing was causing serious erosion was returned with the laconic en-
orsement, "Probably true, but best not to discuss it
now."

Already aware of gullying, Bennett in 1905 came to
understand the insidious and far more widespread
damage of sheet erosion, steadily washing away the
fertile topsoil far faster than it could be restored.
Speaking, writing, and piling up evidence, he was finally able to get permission to publish in 1928 a bulle-
tin entitled "Soil Erosion: A National Menace." Within
five years, during drought in the semiarid West, wind
lifting the soil from land that never should have been
plowed added to the colossal damage done elsewhere
by water. Official opinion could no longer remain indif-
ferent. The Soil Erosion Service, soon to become the
Soil Conservation Service, was launched with Bennett
as chief. Thus the Southland, scene of the most graphic
evidence of faulty land use and management, came to
furnish the advocate of better ways.

Bennett's efforts were made possible by a president
who, in the words of a longtime friend, was "usually
willing to listen and seldom afraid to act," and they
were supported by the first secretary of agriculture to
have a scientific background in his field and by many
enthusiastic public servants and informed citizens.
Further, the public, stunned by what seemed to be economic collapse, was less obstreperous than usual in
opposing change.

Key to action lay in legislation that provided for soil
conservation districts organized by local vote in the
same way that drainage districts had been and fur-

nished with technical advice and service by the federal government. Enabling laws at the state level were gradually passed, not always easily. Experiments were set in motion demonstrating the grave damage taking place under traditional practices. On the whole, the slowest response came from the so-called best farmers on the most fertile land, men who were unaware of the deadly process of sheet erosion. The first, and for a long time to the air traveler most highly visible, efforts at reclamation were in the Southland, that is, in the Southeastern quadrant.

Here rainfall is adequate to support a native forest cover, the growing season is long enough for a maximum variety of crops, and winters generally are mild enough to minimize the need for domestic fuel. Two critical minerals, oil and phosphate, once abundant, have been heavily exploited. The sandy fringe of coastal plain, in many ways ideal for crops, requires heavy doses of mineral nutrients to produce, while the slopes of Appalachia, Ozarkia, and lesser elevations must be handled with care to prevent erosion.

The length, volume, and gradient of the Tennessee River has created both problems and promise. Systematic management finally made possible by a federal authority in 1933 has improved flood control and the regional economy by providing power and at the same time encouraging better land use. Victim of success, the river has been dammed to a degree dangerous to normal flow while the pressure for additional dams continues. Increasing demand for electricity has exceeded the supply of water power. Industry based on this source of energy has expanded to the point where coal-generated electricity has been enlisted.

The result has been a classic economic dilemma. Seeking, as good management dictates, to purchase coal as cheaply as possible, the Tennessee Valley Authority has brought about mining practices in Kentucky and West Virginia that are said to be creating serious, possible irreversible damage. Again, the TVA has brought into sight other clashes of good intention that beset resource management. One example is the conflict between believers in local responsibility and those who put their trust in central authority. Such well-meant differences delayed ministrations of the Soil Conservation Service on the Tennessee Valley. A similar situation arose within the authority as to its basic responsibility, the degree to which power production should be emphasized as against the total impact of river engineering on environment and people. Finally, the natural vegetation of the Southland is forest. Here too the record has been ambivalent. Now that fast-growing pine can be used for paper pulp, it can be harvested for that purpose by thinning in less than two decades. Thus it becomes a crop that can be used on private land to supplement field crops, a practice encouraged by large lumber companies and favorable to a permanent economy.

Florida is the subtropical apex of an inverted triangle whose base extends from New York to Chicago or beyond. The resulting pressure is enormous and the temptation to exploit is in proportion. Although Florida has become highly urbanized, agriculture and horticulture remain major industries. Its sandy soils lie over a limestone base that has retained many lakes and extensive swamps and marshes whose drainage to increase tillage and other activities has had undesirable side effects. Soon after the end of World War I, a considerable area of cypress forest was cleared; a small area reserved by the National Audubon Society remains to demonstrate its unique character.

Symbolic of changed environment was a visit, after fifty-seven years, to what had been the heart of the Big Cypress with its giant trees and rich undergrowth, plant and animal. Where housing, often jerry-built, permitted, cypress trees had sprung up, only to reach a height of some fifteen feet, then die for lack of the water that had been drained away to permit "development." This sparse forest of dead juveniles resembles the stark aftermath of fire.

During the period of this change the population of Florida had risen from less than 1 million to more than 8 million and from a density of fewer than 20 people per square mile to more than 150. A result of this pressure, made possible more by wealth imported than wealth created, has been the replacement of native cover by an assortment of land uses. Depending upon fitness, opportunity, and individual judgments, there has been a varying sequence of living, business, mining, garden, orchard and farming facilities, urban and rural refuges for those escaping from more rigorous climates, and even a degree of industrial development.

The Florida story is not an unbroken litany of environmental abuse. Generous reserves, public and private, have been set aside to protect specimens of its original beauty. Nor should its citizens be especially indicted for its exploitation; rather, blame should be spread over the great triangle we have mentioned. Florida is part of a nation to which its products and facilities are of great value. In attempting to protect their source into the future, the citizens of that nation have a stake. It is reassuring to note that powerful movements to prevent further damage have had strong support from without, as well as from within, what the Spaniards justly named "A Place of Flowers."

6

The Great Pattern

KNOWLEDGE GROWS by many paths. One is the meticulous, detailed examination of a limited part of the whole; another is the scanning of larger and more complex aspects of the challenge to ignorance, in search of pattern. Those who follow these paths lose touch, each with the other, at their peril and that of truth.

The noted physician Dr. William Osler would on occasion stand in the doorway of a hospital ward and give the young students with him his judgment of the illness of each patient, viewed from that point. Then the group would tour the ward as a test, only to find that Dr. Osler's "sight reading" was accurate to an uncanny degree. This was neither trick nor clairvoyance. For years he had followed the course of illnesses, case by case, to the limit of available evidence.

Like the human body, the landscape is an immensely complicated affair. It is at once the scene and product of change, although on a far more deliberate time scale. As with the living body, its changes may be healthy or not so. In either event what is happening can be detected by the eye trained to see. Much of that training could and should be a part of the apprenticeship in living that we, for better or worse, call education.

We smile at the specialist for knowing more and more about less and less, ignoring the spectacle of our restless selves speeding on wheels and wings and learning less and less about more and more. Yet there is comfort to be had in the assurance that knowledge of earth and life can be introduced to the average individual in ways that will give meaning to what he sees around him, wherever he may be. This despite the intuitive differences in skills and interests among people and subject only to the indifference that is a greater threat to our chosen way of life than the rivalry of any foreign power.

Fortunately, although there are none so blind as those who will not to see, there is now open to the rest of us a growing vista into the human past. Just how and when our own species Homo sapiens arose from its ancient humanoid ancestry need not detain us, for within this species today are varieties or subspecies whose common potential far exceeds their physical differences. Assuming that the art of making and using tools is not less than 200,000 years old, and that of domestication not less than 10,000, it is safe to say that human beings, for at least 95 percent of their existence, have been hunters and gatherers. For some groups, notably those living under the harshest of environments, as did the aborigines of Australia and the Seri of the southwestern North American desert, this way of life has persisted by various means of population control and usually by virtue of incredible effort and folk wisdom. No one coveted their territory until modern technology opened up new possibilities. These marginal cultures are not necessarily made up of inferior stock; the Seri, rated as the most primitive of North American aborigines, make carvings of ironwood that rival in beauty and finish those of any craftsmen. And translations from the language of the Sonoran desert reveal genuine poetic sensibility.

It has long been assumed that, since agriculture does make possible more people per unit area, its "invention" was the cause of a sudden revolution in human numbers. This idea is now being questioned. Agriculture did not begin until the habitable earth was fairly well occupied and then it seems to have begun independently in widely separated places. Thus ended, or at least greatly slowed, the classic escape from overcrowding by swarming into vacant lands. Unable to get away, obliged to make the best of whatever was at hand, hunters and gatherers found ways to encourage the growth of plants they had already been using in the wild state.

Certainly in more modern times the pressure of numbers in a limited space has, as in the Netherlands, been accompanied by a remarkable improvement in agriculture. And the centers of crop domestication— wheat in the Near East, millet and rice in the Orient, maize in Central America, root crops in the Southern Pacific region—are all well known. Once such centers had been established, diffusion, based on opportunity and need, must have followed.

Not only does agriculture bring about an economy of space, but an economy of effort, often overlooked because it does entail hard work for those engaged in it. Before 1900, when more than half the population of the United States was rural, it was producing food and fiber for itself, for the rest of our population, and for export, with such commodities as tobacco and material for vintner, brewer and distiller thrown in. Today, thanks to the substitution of fossil energy for human and animal muscle, farming involves less than 10 percent of the total population. But even in much earlier times, agriculture released enough human energy from providing subsistence to develop the arts we call civilized and the urban growth that they foster.

Such activities took place in centers of population, serving the rural areas with markets, records, ritual, manufacture, protection, and doubtless in many other ways. Improved skills and artifacts made possible the expansion of agriculture into areas such as deserts and dense forests where it had at first seemed impossible. Admirable engineering works, accomplished with the simplest of techniques, made possible the growing of irrigated crops in Mesopotamia, the Nile Valley, and the North American Southwest. Conquest of the forests of Denmark before 2,000 B.C. by Neolithic man using stone axes and fire was long a puzzle; such axes in modern hands inevitably broke. Finally, ways were found to mount the stone blade so that the resilient wood of the handle received and cushioned the shock.

With trees thus felled and burned on test plots, the ashes furnished mineral nutrients sufficient for a few years, after which grass and weeds afforded pasture until new clearings for crops were established and woody brush and trees reclaimed the original clearing. This method of land use, known as the milpa system in

Latin America, swidden in Southeast Asia, is still followed in many places. Ultimately, however, it is self-defeating. Where populations become too numerous, the cycle becomes too short to permit recovery; where practiced on sloping lands the soil is eventually lost, as rectangular plots of bare rock visible from the air in northern Mexico testify.

In like fashion, desert irrigation, unless practiced with great restraint and skill, is subject to serious hazards. Its initial success, encouraging population growth, can be self-defeating. One hazard is silting of the ditches that bring water from the surrounding mountains after the latter have been stripped of their forest cover. Photographs from the Middle East show long piles of silt that for centuries have been growing with the annual scooping out to keep the ditches clear.

Another hazard has come from the accumulation of salts caused by the high rate of evaporation in the dry desert air. Like all but a few natural waters, those used in desert irrigation contain dissolved minerals. And unless these are skilfully flushed out, they pile up and render the soil unfit for further crop production.

Again, where the water for desert irrigation comes from underground sources, this water may not now be recharged from the surrounding high lands. With modern machinery it has become possible to pump out water from these sources for agricultural and urban use to the point of depletion. When the end is reached, farms and cities will either have to find some way of importing water or be abandoned.

Yet there is little doubt that the very handicaps of desert regions, challenging human ingenuity at every step, explain why they were early cradles of civilization. The control of water and the recovery of boundaries after flood call for mathematical and engineering skill. The irrigation economy of the Near East has been far more productive than that of the cut, burn, plant, abandon, and repeat cycle of the forested areas of Eurasia.

Between forest and desert lie subhumid to semiarid grasslands, natural homes of grazing and burrowing vegetarians and the carnivores that depend upon them. One of the amazing features of this system, still to be seen in Africa, is the balance between predator and prey, and, as one observer put it, the absence of panic among the latter. Situated as the world's grasslands are, between humid and arid climates, they reflect the balance between precipitation and evaporation. Where this ratio exceeds 1.00, the climate generally favors forest; as one crosses the grasslands toward scrub and desert, it becomes steadily less.

Climatic boundaries are not fixed, but vary from year to year; roughly, wetter than average years tend to come in groups, as do dryer ones. Against these events the natural plant cover is protected by its variety; most species are able to lie dormant during years unfavorable to them, ready to show and thrive at other times. As long as the movement of animals was free, they used it to adjust to variations in their food supply, both seasonally and from year to year.

As early humanity spread over the earth, hunting was the obvious way of life in the grasslands, supplemented by gathering a considerable variety of plants, both those with edible fruits and underground parts. This involved seasonal movements of small bands, following the herds of game, pressure on game increasing with human numbers, improved weaponry, the use of fire, and great "drives" that slaughtered masses of game by forcing them over steep precipices.

In the Old World it became possible to domesticate animals of the steppe by sharing their migrations. In the Americas, except for the use of dog flesh and domestication of llama and alpaca, hunting continued to be the source of animal products, finally exterminating such large mammals as horse and elephant. The abundance of animal life supported by the grasses and other nonwoody plants, along with the vastness of the grassland province and the tough sod with its deep tangle of root fibers, made agriculture both difficult and needless under primitive conditions. So the native prairie persisted in mid-America after the forest east of it had begun to give way, yielding at last to the pressure of human numbers for living space and the use of increasingly powerful tools.

The continents differ so greatly in size, shape, surface features, and location that it is difficult to be very general about the patterns of climate and vegetation that have so profoundly affected human activities. Because of the rotation of the planet, however, there is a tendency for eastern coasts to receive masses of air laden with moisture from the oceans and for this moisture to be lost in moving west and toward the equator. Toward the poles, as temperatures become lower, evaporation is less, and we see, as in Canada, that the grassland tapers out and forest extends to the Pacific coast. Thus twenty inches of average yearly rainfall that permits only desert scrub in Mexico is enough for forest in Canada. And where mountains lie across the path of moving air masses, they take their toll of moisture at the expense of lands lying west and leeward, as do the American Rockies to the Great Basin and southwestern desert.

If one could stretch a curving band, like the diagonal silken ribbon on the portly front of a foreign dignitary, from the spruce-fir-pine-birch of northern Maine to the desert of Baja, California, it would pass through the major zones of temperate vegetation and the climates they express.

Moving southwest across the Lake States, it would cross a growing amount of mixed broadleaf forest, whose beech and sugar maple signal ample moisture. Gradually, as these had replaced the Canadian type of forest, beech and maple diminish, while oak and hickory, less moisture-demanding kinds, take over.

Farther along there would have been, in pre-Columbian time, increasingly large islands of tall grass prairie mingled with the groves of oak and hickory, at length occupying everything save for ribbons of woodland in the valleys. Somewhere near that imaginary but extremely useful line called the 100th meridian west of Greenwich, the taller grasses would have gradually given way to the shorter ones of the High Plains that have held the surface of this semiarid climate against wind and water for millenia.

As our band continued toward the desert peninsula that lies between the Sea of Cortez and the Pacific Ocean, the clumps of grass and other plant life would become farther apart, revealing the bare soil between. Woody scrub, such as sagebrush and creosote bush, would (and still do) signal approach to the desert, bizarre and fascinating to visitors from more humid climes. Here the dominant plants exist in a kind of truce, spacing themselves in a way that insures survival. Mostly, they store the rare and precious rainfall with the help of shallow roots; and so far as possible they discourage destruction by hungry animals through thorns and repellant taste and odor.

Along the pathway we have described there would be many interruptions to its gradual change. Mountains, as we have noted earlier, intercept a sizeable share of the moisture that comes in across them. Thus the stubby Ozarks have been able to support forest in a climate otherwise suited to tallgrass prairie. Yet so delicate is the balance here that the northeast face of Rich Mountain on the border between Arkansas and Oklahoma supports woodland normal to Indiana or Ohio while across the crest facing southwest is a cover of dwarf oaks.

Farther west on more gently rolling ground and flanked by prairie lie the cross timbers of Oklahoma and Texas, described long ago by Washington Irving. These exist not by grace of greater moisture from the air, but because they grow on sandy soil that yields its moisture more generously during dry intervals than do the clay loams of the surrounding prairies. They are noted here despite our need to describe the background of land use as broadly and simply as possible to remind us that the impact of climate can be cushioned or magnified by textures of soil and vegetation as well as differences in surface elevation and exposure.

It can also be influenced by the chemical makeup of local soils. To take two extreme examples, alkaline salts accumulate in arid climates where poor surface drainage combines with high evaporating power of the air, while strong bones and powerful animal bodies develop where, as in the Blue Grass region of Kentucky, soil comes from the weathering of calcium phosphate. Elsewhere native animals have in the past distributed nutrient materials by moving about and making up the lack of needed minerals by visits to salt licks, some of which preserve the remains of animals now extinct.

Concerning the animal life that has varied along with the types of vegetation that have supported it, suffice to say that in the orderly processes of nature without human interference, the activities of plants, animals, and microorganisms are close-knit. As we shall see, where modern practices break this bond, man is increasing the hazard to future generations. Meanwhile we shall assume that the two-billion-year record of life on earth may have some lessons for us in our effort to survive and make living worthwhile. To that end let us briefly examine process among living organisms.

Thoughtful scholars have said that there are no laws of history. This is half-truth, for history is process and there are rules of experience that govern process. Back of the infinite variety of possible human choices and behavior lie the patterns of energy and material change that involve dynamic units, be they physical particles or individual living organisms, the human being included.

Whether we watch a crystal of colored, soluble material dropped into a glass of water or observe a busload of strangers embarked on a long trip, we see process going on. Color tends to become uniformly spread through the water, revealing an approach to balance between the two kinds of particles involved. Passengers on the bus begin to thaw out, communicate, and assume identities and roles, whether those of sage, clown, nuisance, or simply that of agreeable companion. For unlike the particles of crystal and water, each passenger, although a dynamic unit, differs in significant ways from all of the others. Nevertheless, both situations tend to move toward some kind of system or balance, subject always to interference, internal or from without. If, for example, the water were frozen before the crystal had dissolved, the process of diffusion would be arrested at whatever stage it had reached, resuming if and when melting occurred.

At a far more deliberate rate, this same tendency to work *toward* a balance takes place as plant and animal life move onto an abandoned roadway or ploughed field, a sandbar, exposed rock, or within a lake that is gradually filling up with minerals and waste materials.

Just as nature abhors a vacuum, it may be said that she will not tolerate idle surface on the earth. Whenever a new area is added to a continent in any fashion, the slow and endless process of covering it begins. Whether this new surface is rock, sand, or clay, there are venturesome kinds of plants, often very small and insignificant in appearance, that form the advance guard of nobler kinds. These pioneers may look like scales or tufts of living green, or, if the new surface is not solid rock, they may be our hardier and more familiar weeds. Whatever their type, they slowly change the conditions of light, moisture, chemistry, and structure in the place where they grow. Like the true pioneer of the human race, they do not thrive under the crowded conditions that their own activity has produced. In consequence, they give way to other more tolerant kinds of plant life and the process continues.

At each stage there are appropriate types of animal life that fit into the growing community and are a necessary part of it. In addition there are many invisible forms of life whose activity is just as essential to orderly progress as that of the more conspicuous plants and animals. As time goes on, the remains of dead organisms accumulate and by the action of the invisible bacteria are built into the growing soil, enriching it chemically and improving its physical texture. From this cource comes the dark material known as *humus,* characteristic of rich and fertile soil.

The soil begins to assume a marked and definite character, depending partly on the kind of raw material at hand for its construction. But in even larger measure the character of the soil is stamped by the climate under which it forms. In the forests near the coast the abundant rainfall seeps down through the thin, rich humus left by the forest leaves. As it descends, it extracts from the soil great quantities of lime and other such soluble materials. These later reappear in the springs and underground streams of hard water which are so common a feature in nature. These soluble materials are mostly alkaline. The material that is left is often acid, and so is, frequently, the humus that is on top. This is easy to understand if one will remember how ensilage and kraut, when allowed to stand, sour through the formation of acids. The soil at the other extreme of climate, namely the semiarid and arid regions, receives a relatively small amount of plant and animal material; this tends to dry rather than to ferment as in the forest regions. So little rain falls and so dry is the air that what water there is in the soil tends to be drawn to the surface, bringing with it the dissolved salts of lime, magnesium, and other alkaline material. In consequence, we often see in the drier parts of the country soils that are alkaline instead of acid. In many instances this alkali is not excessive in amount. It then serves to provide the abundance of needed minerals that enable the desert to blossom forth if water is supplied by irrigation, or whenever a brief, infrequent rainy season occurs.

Between the desert and the forest, in subhumid grasslands, occur soil changes of the utmost importance to mankind. Here the abundant fibrous roots of the grass penetrate deeply into the soil and rot in place when they die. The dead leaves of the grass form a mulch, or blanket, that equalizes the movement of the water. There is neither a surplus of evaporation nor any violent percolation. Runoff and scouring are reduced to a minimum. The result is a layer of deep, rich, black soil of great fertility, neither excessively acid nor extremely alkaline. Besides the grasses, there are many wild legumes such as lupines, vetches, and clovers, on whose roots are found bacteria that fix the nitrogen in the air into substances that greatly enrich the soil. The wastes and remains of the many grassland animals also add such materials as phosphorus, which is of the utmost importance for plants.

Through the spectrum of climates, from moist to dry, from forest across grassland to desert, this process of slow development has gone on and, wherever renewed, still proceeds until some kind of balance is reached. Whether and at what level further change may be arrested or even reversed depends upon such local conditions as soil character, exposure, fire, drainage changes, or in broader perspective, climatic change.

Useful and not wholly theoretical is the concept of a balanced condition known as a *climax* community of plant and animal life. Here, where limiting factors are at a minimum, the kinds of plants and animals present are delicately adjusted to each other and to surrounding conditions. Barring major changes such as a new pattern of temperatures or moisture, geological episodes such as volcanic or glacial activity, submergence, etc., or biological invasions (including human), the climax, in principle, tends to perpetuate itself. The young, unlike those of weeds and other pioneers, can thrive under conditions favorable to their adults, given only the space vacated by deaths among the latter.

It is not too far off to say that when life enters new territory, environment dictates the terms. In effect, the first comers must be able to put up with such extreme conditions as untempered exposure to light and wind, excessive moisture or dryness, and raw-mineral footing. As time goes on, new species replace the original, working relations are established within the increasing variety present, and in the mutual relationship between environment and organisms, the latter play a larger part than did the pioneers that preceded them. The result is a working pattern not inaptly termed an *ecosystem,* an expression of the interaction that exists not only among organisms, but between them and their surroundings.

What happens, in terms of the energy and materials involved in this process, is an approach to a kind of equilibrium known as a *steady state*. Since it is kept going by solar energy that builds up plant materials and is stored in them, the system is, in terms of energy, an *open steady state*. But the materials (organic compounds) are broken down in use by plants and animals and eventually returned to air and soil to be used again. So in terms of material change, the steady state is a closed cycle, or a *closed state*.

Central to these remarkable processes is the chemical element known as *carbon,* which, strange to say, too seldom receives the attention it deserves in beginning chemistry. Carbon units have the ability to combine with themselves and a variety of other elements, making possible an enormous number of compounds, including the familiar sugars, proteins, and fats. Energy is needed to build these compounds; it is stored in them and released when they are broken down. At times in the geological past, this breaking down has been delayed, piling up the residues of organic activity in such forms as oil, gas, and coal, which we are now engaged in breaking down at a rate incalculably faster than they accumulated. Used to operate heavy farm machinery, more of their stored energy may go into an acre than is returned in the harvested crop!

Less spectacular, but more essential to human survival, is the accumulation of carbon compounds in the soil and the activity this makes possible. The wastes and residues of plant and animal life, over and beyond those released as water vapor and the gas carbon dioxide, accumulate as dark, jellylike or gluelike stuff called humus. Humus not only returns nutrients to the soil, but profoundly affects its structure and activity. This is so true that the popular notion that the darker the soil the richer is usually correct.

Humus not only supports the small forms of plant and animal life that work within the soil, but forms the mineral particles into "crumbs" necessary to the economy of air and water. These crumbs also serve as a kind of banking system for soluble plant nutrients that otherwise would be washed away, paying them out to plant roots as needed.

Just as a house becomes a home, so the rock particles of the earth's surface become soil by being lived in. Soil is more than a mere mixture; it is a system whose nature, revealed by a vertical cut, is expressed as a profile showing the layer-cake relationship of parent inorganic mineral material to organic humus. This profile is basic to proper use and care of soils in the great zones of climate and vegetation. Failure here has been far from the least of human tragedies.

In the humid climates that favor forests, their litter accumulates with fallen leaves, whether these are needles or broad blades. The waste from the aromatic needles of pine, spruce, and fir does little to prepare a fertile farming soil, but occasional fires permit the growth of aspen or other broad-leaved trees that do provide mellow humus. On the whole, where cone-bearing "evergreens" thrive, the world can well use the wood they provide.

The broad leaves of oaks, maples, and many other deciduous trees shed each autumn, tend to form a rich, damp blanket above the mineral sands, silts, or clays on which they fall. Worms, insects, microorganisms all thrive here, breaking down the annual tribute that sustains them into simple forms that can be used again. The annual excess of rain and snowfall over evaporation tends to leach out this surface layer as the hot water in a drip coffeepot does with the broken grains through which it drops. The residue tends to be acid and the water that soaks away tends to carry lime and other minerals into springs, underground waters, and streams.

The soils that develop under forest, comprising at least the eastern third of the United States and most of northern Europe, are fabulously rich as "new ground" for a few years after being cleared. Thereafter, they become less fertile, usually because the thin layer of forest humus soon washes away or decays. Unless converted skillfully to pasture and legume rotation, or enriched by a constant supply of animal and plant wastes, their agricultural value declines rapidly, to be maintained only by supplies of artificial fertilizer that does little or nothing to restore crumb structure. If one were obliged to form a ready judgment on land use where broadleaved forests once grew, it would not be too different from that which has been worked out blindly, but would suggest far more responsible and enlightened management.

Rainfall being adequate for domestic and other uses, the problem is to conserve water quality. For towns of moderate size, enough falls within urban limits (30 inches or more) to serve normal consumption, if it could be kept clean and stored in cisterns. Gardens, dairies, and orchards, suited to rather intensive care, are possible. Fertilizers would be necessary, but could be kept at a minimum if animal industry could be kept going instead of totally mechanized crop production. Woodlots of valuable hardwoods could insure wildlife crops as well as needed materials and facilities for recreation. And because population is often denser where water is abundant, care should be taken not to waste land by failing to put it to the best possible use, so far as that can be foreseen. Ribbons of concrete and masses of masonry have much in common with lava flows, taking over land regardless of its potential.

As the humid climate suitable to forest passes desertward into subhumid conditions, trees give way to tall grasses and many kinds of nonwoody flowering plants. (These latter are called forbs by some, flerbs— probably a condensation of "flowering herbs"—by others; only forb has made it into the dictionary.) Compared to forest, biological activity in the subhumid prairies increases underground, where the dense turf of fibrous roots and such storage organs as bulbs support abundant animal, fungous, and bacterial life. Here humus accumulates to depths of three to five feet or more, holding and recycling mineral nutrients and providing the most productive of agricultural soils.

Aboveground the leafy shoots that originally sustained vast herds of grazing animals and their hosts of dependents form a kind of dry, matted litter as they die, locking up considerable nutrient matter. This slows down growth until the occasional fires set by lightning consume the litter and release the nutrients in its ashes for reuse. But too frequent fires, such as those often set by man, as well as overgrazing and other abuses, reduce the surface cover and expose the rich black, humus-laden soil to the devouring action of wind and water.

This area, vital to food production, is at present in difficulties. Land prices, based on possible "development" for uses other than farming, are inflated beyond any relation to income from crop yield. This is reflected not only in real estate taxes, but on estate levies that amount to confiscation. Animal industry has become increasingly one of finishing range cattle in huge feed-lots where organic wastes, no longer available on the average farm, are instead a disposal problem.

Production costs of food and feed crops multiply while human effort per capita per acre diminishes with the use of heavy machinery and chemical agents, many of them from petrochemical industry. The depth of the original dark humus layer has masked, until recently at least, its steady loss through erosion. In short, modern practice has little if any relation to the processes that have built up and maintained the world's granary. Meanwhile, the combination on the same farmstead of plant, animal, and human industry of the best traditional agriculture, which might have preserved the essentials of natural processes, has been displaced by lavish, energy-consuming mass production methods known as agribusiness.

Beyond the tallgrass prairie and its transition to the semiarid shortgrass climate, the results of highly mechanized one-crop farming were the occasion of the first edition of this book and have been sufficiently discussed in other chapters. There have, of course, been improvements in techniques and concern, but the long-range effects are still not promising.

Within the shortgrass country, the grim lessons of the Dust Bowl are easily forgotten. Less than twenty years after that event, a dry-country wheat farmer was overheard to boast that he had cleaned up and that "she could blow all to Hell now" so far as he was concerned. Areas of shelterbelt, planted to temper the winds, and even of streamside woodland have been cleared for plowland, greatly increasing the danger of erosion.

Encouraging this trend, countless wells have tapped underground water, regardless of whether it is being recharged as fast as used; generally levels are dropping rapidly, as much as six feet a year in places. In much of the southwestern desert there is no recharge. The water being pumped out is correctly called fossil water, having been stored during wetter glacial times, more than ten thousand years ago. Once used it is gone, as much a nonrenewable resource as coal, oil, or gas. At first this harvest from what Charles Bowden has called "The Killing of the Hidden Waters" was used chiefly to expand irrigated desert farming. But as the attractions of the Sunshine Belt for winter refugees has increased, it has been necessary for urban areas to purchase the water rights of as many farmers as possible—a temporary safeguard at best.

7

Only God
Can Make a Tree

IN 1913, FIVE YEARS before he was killed in battle, Joyce Kilmer published "Trees." Often quoted, this poem became legendary, one story being that it was written as a final message from the field the night before the author's death! But what concerns us here is that "Trees," whatever its merits, presents only one of many approaches to a form of life of enormous significance to man.

The lumberman looks upon the tree with an appraising eye, in terms of board feet. The artist views it as a glorious blend of symmetry, light, shade, texture, and color. The scientist for his part sees it as a marvelous, intricate laboratory whose workings are a perpetual challenge. Actually, mankind has never viewed the forest with the single eye of Scripture. For while it supplies many of his urgent needs, it is his rival for space with which to supply other needs, notably the production of food. Less obvious is the vital role of forest in the economy of soil, water, and air.

Cortez in 1519 looked down from a mountain pass

upon a basin of many lakes surrounded by forested hills. Even the tough soldier Bernal Diaz was awed by the magnificence below, the product of a civilization nourished by a remarkable system of gardens. Although the Basin of Mexico had been flooded occasionally before the Spanish conquest, floods became more frequent and severe thereafter—no fewer than five between 1519 and 1607. Along with water came heavy loads of mud.

With a beautiful display of the kind of impromptu policy against which the late Arthur Morgan protested to the end of his long life, the Spanish government ordered the lakes of the basin to be drained. To accomplish this, they chose probably the ablest engineer at hand, Enrico Martinez, said to be Dutch and of Jewish descent.

The Netherlands had long been preeminent in skilled use of the land, as were the Moors and Jews in scholarship, so it is not surprising that Martinez diagnosed the trouble with a clearer vision than those who ordered him to remedy it. In 1606 he wrote that the once-forested hills around the basin had been *"descarnados"* (stripped to the bone) by the cutting for lumber and fuel and by the subsequent heavy grazing of cattle *"de los Cristianos."* Many trees, we are told by others, were victims of Spaniards "homesick for the treeless plains of Castille."

Not to be forgotten is the ancient and insistent need of the human race for fuel, to this day a major use of wood, especially but by no means exclusively where fossil fuels are not to be had. Even the stone axes of Neolithic times, hafted and used properly, can be surprisingly efficient in felling trees, as Danish experiments have shown. With the invention of steel, an alloy of carbon and iron that requires heat in the making, assaults on the world's forests were multiplied. Steel tools speeded up the cutting and working of wood, while demands upon wood for the making of charcoal and production of weaponry were endless. By the time of Charlemagne, vast areas of European forest had been cleared.

To its first European settlers, eastern North America seemed an almost unbroken forest. By 1825 some three-fourths of New England had been cleared for farming and other uses. With the opening of the Erie Canal, competition from the more fertile soils farther west led to abandonment of farmlands and return of the woods in New England. Today proportions are almost reversed, with at least three-fourths of some New England states now forested. In comparison, the rich soils of Ohio had an original forest cover of more than 90 percent; forest remains today on less than one-fifth of that state.

In medieval Europe feudal landowners often set aside extensive hunting preserves, thus preventing destruction of forests. The gathering of fagots of dead and down wood, a privilege granted to the poor, doubtless helped keep alive an appreciation of the importance of forests until political change made possible the establishment of community forests in western Europe. In the Balkans, place-names preserve the record of vanished forests of beech and oak; here and elsewhere, reforestation has suffered from a fuel hunger that does not spare young planted trees. In one-time forested Greece, the visitor learns that wood finish is a luxury reserved for special rooms. Elsewhere marble must be made to do.

By the mid-nineteenth century, Germany, Switzerland, and Scandinavia had forestry programs that might, if adopted in the United States, have prevented colossal waste and done much to cushion our present economic distress. But the decades following the Civil War were times of pillage unrestricted. Between 1850 and 1900 the center of lumbering shifted from western New York state into the lake states and beyond. A contract dated soon after 1900 called for "clear Michigan White Pine not to cost more than twenty dollars a thousand." Wood from old railroad cars, still usable for sidewalks and outbuildings, was often burned to salvage the metal in it. Building plans, advertised in a catalog dated April, 1875, were as obviously devices for sales promotion of unnecessary lumber as modern magazines are for the sales of liquor, cigarettes, automobiles, and other items whose rate of production treads hard upon, indeed often overrides, the heels of actual need.

Had the United States in 1880 seen fit to adopt a policy already long accepted in Denmark, for instance, we would today be harvesting timber almost a century old. Today's contractors would not be able to say, as one did recently, that for the first time in his experience, materials were costing more than labor! So ruthless was the exploitation of these rich, virgin stands of conifers and hardwoods that cutting was often followed by fire. Not only was no provision made for leaving good seed trees to insure replacement, but often the ground was left in such condition as to delay indefinitely the return of forest. In an attempt to squeeze the last drop of gain, settlers were encouraged to move in and attempt farming where there were no alternatives to failure. Fortunately, in recent years, some state governments have been able to acquire cutover land and, by a policy of combining recreation with reforestation, increase the possibility of future benefit.

Not all tree cutting was based on the desire for quick profit regardless of consequences. Pioneers in wooded areas needed not only fuel and building material, but also cleared space to grow the food and fiber required by their families, although this often involved burning piles of huge logs that would be today extremely valuable. At first, warnings of the dangers of excessive destruction were largely intuitive and aesthetic, at times combined, as they were in Thoreau, with scientific understanding. From an early and emphatic protest by George Perkins Marsh in 1864, the growing concern of naturalists, tree lovers, public-spirited citizens, and even a few woodworking industries finally became effective at the insistence of the first President Roosevelt.

Although some of his predecessors had crossed the continent, Theodore Roosevelt was the first of our presidents to know the Great West in an intimate way. With the help of foresters Pinchot and Greeley, he added to the national forest reserves nearly three times as much as had Presidents Harrison and Cleveland together. Congress had passed antitrust legislation in 1890, but it was still reluctant to curb exploitation posing as enterprise, lowering the boom on Roosevelt before he could do more. Yet thanks to him, forest conservation became a recognized feature of our public policy. The Forest Service was built into an active arm of government, destined to be the object of a tug of war between Interior and Agriculture where it remains, and to encounter pressures from various special-interest groups. Under these often difficult circumstances the forest service personnel, recruited largely from individuals who have a genuine appreciation of trees as well as their products, have shown consistent patience and tact.

Meanwhile the public interest created after 1900 was eclipsed by two great wars that, like all wars, were insatiable devourers of natural resources, not the least of which was timber. Separating them came the disasters of the 1930's, reawakening conscience between slumbers. Steadily the era of much-for-little was drawing to a close; business was being forced by circumstances to be businesslike. The lumber industry and associated enterprises were becoming corporate, with administration responsible to stockholders. Operation could no longer be a matter of cut and run; serious attention had to be given to maintaining future supplies of raw material. As corporate policy developed, government cooperation became available— and welcome—for both large and small owners of woodland.

Roughly speaking, about one-third of the United States is considered to be forest land, regardless of quality, condition, and promise. Of this, more than 70 percent is privately owned, a little more than 20 percent federally, and a remaining less than 10 percent is in the hands of state and smaller political units. Consequently, the management of privately owned forest is a matter of the greatest concern, and the shift to responsible corporate management a welcome change.

However, business operations are subject to many hazards, including bad judgment and pressures that force liquidation. The drive for maximum profit in a system of free enterprise that in the past has led to heavy exports of oil, phosphates, metal, and machine tools—sometimes used in the armament of future enemies as well as by friendly countries—also involves wood. Our chief exports of this commodity are rough lumber and wood pulp, the latter being one of our chief imports as well. Considering the exorbitant increase in the cost of paper and building material, a wise and benevolent despotism would probably decree a halt in exports until production balanced our domestic needs. Having seen the cost of lumber increase more than tenfold in my lifetime, I can scarcely be blamed for this mental excursion in political futility.

The exigencies of business involve yet another hazard to wood production on privately owned land. *Exploitation* was once a respectable technical term when applied to mineral deposits, not a synonym for piracy as it is today. As far as land is concerned, *development* has taken its place, highly regarded as a source of quick and considerable profit, advancing at the expense of much land that, again invoking our benevolent despot, would serve the future better if devoted to agriculture or forestry.

Vital as the national forests are to a sustained economy, neither their size nor character is enough to provide what is needed. Their acquisition was not begun soon enough to give government first choice everywhere. By no means all of it is prime timber land; even where it is, there may be other sources of wealth to tempt the enterprising. Grazing and browsing, mining, hunting, fishing, and camping all compete for place on the program of the national forests. To one with more than a perfunctory interest in and admiration for professional foresters, it was disturbing to see the troubled but conscientious reaction of these already burdened individuals when commanded by government to become responsible for providing recreation. The most recent budget granted to these public servants favors timber marketing and public enjoyment, while skimping the allowance for future timber production.

Fortunately the schools of forestry turn out a product equipped to deal with a wide range of problems. The demand for foresters, like that for any professionals, has its ups and downs. The record of individuals trained in forestry but obliged to find other employment is a remarkable one, ranging into business, engineering, science, and administration. Any failure of the Forest Service to measure up to its promise is far more likely caused by government and public than by service personnel.

One of the basic problems of the national forests, as of the public domain administered by the Bureau of Land Management, is to develop a reliable record of land-use potential, i.e., the best possible use to which any given land could be put. Without such information any long-range policy becomes as impossible as the solvency of a shopkeeper ignorant of his own wares. The first surveyors of the Midwest knew trees and often used them in their notes as indicators of promise.

Any ranger, given time for it, is likely to have a good notion of what would be best for the various sites in his charge.

To date, unless available information is in error, no systematic and thorough land-use-potential survey has been possible. In contrast to western Europe, where one owner of some hundreds of acres of forest keeps fifty men busy year-round, remarking that he would not do so if it were not to his advantage, it is doubtful whether a similar area of our national forest would be in charge of more than a single trained worker. To this must be added lack of official interest and support and occasional covert opposition with its own notion of what for.

At best, complications are many. Topography, soil quality, and moisture supply often vary greatly within short distances. Much "forest" is treeless, some for natural reasons, not a little, like the fringes of modern deserts, having grown by human action. A long stretch of the Front Range of the Rocky Mountains still bears marks of fire set by a hostile Indian leader shortly after the Civil War. Farther south, a considerable area of one of the national forests looks as though it has always been grassland; but the trained eye can still find evidence of an early, extensive forest fire. Word of mouth credits this event to somebody who desired to increase grazing conditions on everybody's land during the 1880's. Whether this rumor is true or not, there is no doubt that users of the range today are complaining that clumps of spruce that managed to survive are encroaching on the grassland!

Among the many burdens of professionals in charge of natural resources is the recurring effort to decide, by

public opinion and legislation, matters that call for
competent technical expertise. Only by patient education of sportsmen (using here "men" to include, as is
the usage among biologists, human beings of both
sexes) have trained wildlife managers gained the right
to set seasons and limits, a privilege previously in the
hands of state legislators.

Rightly enough, the public is concerned about fires
that consume vegetation, be it forest, woody scrub, or
grassland. Sad to say, many of the most destructive
have been caused by human carelessness, even malice,
and for these there is no excuse. But in nature, fires set
by lightning are a part of the normal regime. Heavy
growths of plant life tend to bottle up the soil nutrients, thus slowing further growth. These unavailable
materials accumulate in the dry, dead waste of scrub
and grassland and in the living and dead stuff of the
forest. Natural fires, usually occurring at infrequent
intervals, return nutrients to the soil in the form of
ash, thus renewing the cycle of growth.

We hear frequently of the costly damage to housing
in the woody scrub known as chaparral in California.
Competent scientists, who seem not to be heeded, know
that the trouble is caused primarily by the *prevention*
of fires for long periods of time during which the dead
branches and leaves accumulate, furnishing masses of
kindling whose burning, once started, is almost impossible to control.

Some kinds of valuable trees, notably the redwoods
of the West and the pines of southeastern United
States, are less injured by occasional ground fires than
by the development of heavy undergrowth. For them,
and in certain conditions for other kinds of plant life,

the judicious use of fire is a part of proper management. Clearly the role of fire, like the nurture of desirable wildlife and many other biological problems, is a challenge not simply to good intentions, but also to expert professional knowledge.

Whatever substitutes may be devised, our demand for wood must continue. To produce such substitutes requires the use of energy, mostly from fossil fuel, product of ancient sunshine. Wood, in contrast, is being produced by living plants, using current radiation from the sun. Here we have the privilege, not ours in the making of oil and coal, of influencing the amount available. Back of any techniques, clever as they may be, is the will to use them. This in turn depends upon the intangible but none the less real values of our culture.

These values are the distillate of experience, the expression of struggle. As far as natural resources are concerned, the issue can be clearly drawn: maximum immediate advantage versus future assurance, tempered of course by provision for genuine present need. It has been truly said that the only consistent conservatives may well be the conservationists.

Far more profound than the stark economics of the marketplace is the brief for prudent management of forest resources. While producing an essential commodity, the trees of the forest are a powerful factor in maintaining a favorable oxygen balance in the atmosphere. Cushioning the harsh impact of climate, they shield and stabilize the soil as well as contribute to its development. The spongy litter of the forest floor serves to restrain and regulate the flow of water; since the days of classic Greece, it has been noted that when forests are felled, springs dry up and floods become rampant.

Finally, however sound the bill of particulars for forest utility, the facts of political life show that cold logic has tough sledding unless reinforced with a measure of emotional warmth. Granting that we must first cherish in order to protect and that grassland and desert have their appeal, tree and forest alike speak their message of beauty to the seeing eye. And to rich aesthetic appeal they add, for the thoughtful and compassionate, the ethical value of whatever is good for those who come after we are gone.

8

Leaves of Grass

IF THERE WERE NO GRASSES, man would be just another animal. Only the grasses, whose leaves grow from the base instead of at the tip, can sustain the continued nibbling of grazing beasts; to a degree, they actually thrive on it. Primitive man, emerging from forest, found open country thronged with roving herds. These enabled him to change from gatherer to hunter. Later on he learned to tame the more useful kinds of grazing animals, at first following, then as herdsman guiding, their seasonal movements.

Again, by the domestication of certain highly favorable grasses man found bread, respite from unceasing movement, and ultimately civilization, for the first great centers of urban culture were tied to the culture of maize, rice, wheat, and other cereal grasses. Whether agriculture was the "cause" of human crowding or a necessary response to it need not concern us here. We can be certain that without farming to feed them, cities could not exist.

Despite the part played by grasses in the development of human cultures, we have accorded dismal treatment to our grazing lands and the soil upon which our bread must grow. Nor is our debt for food alone. The great grasslands, apart from the abundant wealth they yield, are the strategic buffers between civilization and the desert. With notable exceptions, the great centers of population are in regions whose climates would, if given a chance, produce forest. From there, human numbers shade out to virtual invisibility in the desert, save where technology and outer wealth intervene. Thus we pass from Massachusetts, with more than eight hundred people per square mile, into Oklahoma with more than thirty-seven, on through Arizona with some fifteen in 1970, into Nevada with fewer than five. If numbers mean anything, it is not good to have deserts grow in size. And against that growth our best defense is the tight-knit turf of the grasslands. If such is the situation, it is well to know more about it.

Unlike trees, grasses have not long had friends who were organized into potent groups. Though far too few for the services needed, we have thousands of foresters. But scientists who understand the technical problems of the grasslands can be numbered in the hundreds at most, and these are largely concerned with grazing and range management. Of course, the reason is not hard to understand. When a tree is cut, it leaves "a hole in the sky." When forests are destroyed, everyone is aware of that fact. Even when the forest deteriorates, one does not need to be a trained forester to sense what is taking place. And when the price of lumber goes out of bounds, the situation speaks for itself, so greatly do we need the commercial products of the forest.

Grass, on the other hand, is to most people simply a green carpet. The kinds of plants, their abundance, and their vitality are matters that escape the casual. As long as all is green, all is well. To the untrained traveler, the great grasslands are a matter of indifference or even an unutterable bore as he moves through them in his travels.

A waggish professor in Salina, Kansas, used to assure his classes that if current eastern opinion were to be trusted, they were sitting on the eastern boundary of the Great American Desert. That term was not coined for desert at all, but for the land that was to Willa Cather, author of *My Antonia,* alive and glowing with beauty:

As I looked about me, I felt that the grass was the country, as the water is the sea. The red of the grass made all the great prairie the color of winestains, or of certain seaweeds when they are first washed up. And there was so much motion in it; the whole country seemed, somehow, to be running.

The grass affects our pocketbooks through the operation of a maze of physiology, technology, and economics, upon which many other factors impinge. The result is that when the price of meat soars too high because of a failure of pasture, we express our resentment not by organizing to conserve the grasslands, but by voting out the old officeholder and voting in the new. As immediate logic this is, of course, execrable, but it may not be so bad in the long run if the politicians are shrewd enough to see what really lies back of the trouble. The terrific dust storms that during the 1930's and again more recently choked nostrils and dimmed the sky may be worth their cost if they focus our attention upon the grasslands, so that we may understand indelibly that they are as vital to our own highly developed civilization as they ever were to Abraham, the lord of flocks and herds, dwelling in his tent.

The Europeans who first explored America came of a stock in whose tradition all influence of a steppe or grassland environment had long since disappeared. Their first glimpses of great expanses of natural grass, or "prairie" as they called it, amazed them. And on through to very recent times the explanation of this kingdom without trees has intrigued and scientific and popular imagination. There is really an imposing list of theories as to the "cause" of the prairies—fire, soil, grazing, as well as climate having been suggested. All of these, and other influences as well, may play a part under appropriate conditions, but we know today that the principal factor is climate. As a rough and ready expression of this, forests tend to occur where there is a greater annual fall of rain in inches than the air, on the average, will draw back in evaporation. Where the reverse is true, grassland occurs, or if the evaporation is still more intense, scrub and desert. Refined studies of the relationship between rainfall, evaporation, and temperature, as well as their seasonal pattern, give us a map of North American climate showing remarkable similarity to the vegetation map.

We have said that the average person looks on grass as merely a green carpet, studded, on occasion, with flowers. But just as the loving eye of the connoisseur, roaming frequently over his treasured Persian weaves, will detect new beauties of design within design, so it is with the man who knows the grassland. For him it is a pattern of infinite richness. So overwhelming is its variety that long years of patient study have been required to see order and law behind the apparent chaos. Yet certainly order and law are there.

From the infinitude of possible sounds that strike the ear, some eighty-odd intervals of pitch comprise the keyboard of the piano. Give a musician these and he will evolve a wilderness of beauty. There are thousands of kinds of plants found in the grass country, including many legumes that enrich the soil by fixing nitrogen from the air with the aid of bacteria. But the nature of particular kinds of grassland is revealed by the way a relatively few dominant species are deployed to fit the vagaries of earth and atmosphere.

The grasses themselves fall conveniently into three groups, according to height—tall, medium, and short—with roots whose depths are in proportion to their tops. These are distributed according to the abundance of available moisture, whether it be conditioned by climate, soil, or topography. Just as the maize of the Iowa farmer is taller than he, while that of the Hopi Indian is scarcely knee-high, so it is with the native grasses. The tall kinds flourish in the humid eastern prairies, the short in the dry western plains, with the medium between. Or if, in any part of the grasslands, conditions of soil and topography produce sufficient degrees of difference in the moisture supply, the various types of grass may be found growing not far apart, each in its appropriate place.

Even under uniform conditions there is usually a mixture. Then in wetter years the taller grasses dominate the scene, in drier periods the shorter kinds flourish best. So flexible is this battery of plant life, and so profound its resources for meeting all of the possible vicissitudes of its habitat, that it is seldom caught off guard. Unlike the fields planted by man, where the outcome is staked on a single kind of crop, there is little or no chance for complete failure. In a particular year cultivated crops may bring much richer returns than native grass, of course, and often do so. But year in and year out the yield of the grassland is sure—that of the planted fields, never. The crops are like speculative securities, the wild grassland like government bonds of a nation whose pledge is sacred. No accountant is needed to analyze the kind of enterprise into which American grassland agriculture has let itself. Instead of there being an ample reserve of native grass on each farm, it is difficult, increasingly so, to find areas that are suitable for study. In the early 1930's two botanists reported that they had been obliged to travel thirty thousand miles in their search for adequate specimens of what seemed then to be a kind of plant community on its way to join the dodo.

Meanwhile, although the deep black A-horizon or topsoil of the tallgrass prairie continues to tempt the plowman, considerable areas have been set aside as preserves, notably in Kansas. Elsewhere, prairies have been rebuilt by planting species once present. This has been done, for example, in Wisconsin and southern Ohio.

Respect for the prairie as an asset in itself rose

where, during the protracted drought of the 1930's, acreages that had somehow escaped the plow were the only productive islands in a sea of blowing dust. The drier kinds of grassland, occupied by short and medium grasses, have suffered less than the humid, tallgrass province from the direct inroads of the plow, although their increasing destruction from this source has been a major factor in serious, widespread and recurring dust storms. What has been said of the speculative character of ordinary farming applies even more strongly to semiarid than to the moister areas.

The plow, although the most thoroughgoing and irrevocable, is not the only agent of change. With it must be considered two other influences that have ever been operative in the grasslands, but whose effect, until the advent of the white man, was on the whole moderate and beneficent. Fire and grazing are a part of the normal experience to which natural grasslands are adjusted, along with the fluctuations of climate and the variations of soil and landscape. Unless utilized fully by normal grazing, the tall and medium grasses produce a surplus of dry growth that rapidly accumulates, insulating the soil from the warming rays of the sun, locking up needed nutrients, and obstructing the growth of new seasons. Because the air is drier than in forest regions, this surplus of dead material aboveground does not readily rot down to form humus in the soil. Under such conditions it appears that fire at infrequent intervals may actually be of service, with benefits outweighing any damage it might cause. Certainly there is plenty of evidence of such fires, not only those set by Indians before the white man came, but even those of geologic time, when atmospheric causes such as lightning or meteorites furnished the torch.

The very fact that both grazing and fire can be beneficial has made the problem of their rational control under crowded modern conditions a very difficult one. And their evil effects are the more sinister because, as explained in an earlier chapter, they are not immediately apparent. Persistent overgrazing and repeated burning inaugurate not prompt desolation, but instead a slow insidious change in composition, ultimately reducing the total yield as well as the food value. Stock-poisoning trouble more often than not starts with overgrazing. There is more room for poisonous plants to become established and more temptation for the stock to eat them in the absence of highly palatable forage. Thorns and spines enable the plants that have them to persist in spite of overgrazing, and ultimately to spread, thus increasing the numbers of those inedible plants.

The plants of the virgin sod are mostly perennials, coming up year after year from the same underground systems. Once established, they depend much less on seeds than on new buds for their spread. This in fact is one of the most important features of the marvelous flexibility of the grasslands that we have mentioned. Each individual clump is prepared to put forth few or many new shoots each year, as conditions happen to permit, or even to suspend operations for a time and live upon its underground store. Under too-frequent attacks from fire or hungry teeth the clumps of grass cannot hold their own. Shrinking in size, they leave space for brash, short-lived annual weeds of little food value. These may be followed in turn by more persistent and even more useless perennial weeds, such as the poisonous or thorny types mentioned above. All too often, if the abuse is continued, nothing can come into the tortuous spaces between the clumps in time to prevent the elements from scouring them into winding channels. It is not uncommon to find a once-beautiful and uniform sod, that has been reduced to isolated tufts or bunches, each standing several inches above the eroded bare surfaces that separate them. There is a kind of grass that finally comes into these naked spaces after the topsoil is gone and erosion has done its worst: the natives call it "poverty grass." Its dirty, pale gray will show in patches across the landscape for miles, a warning to any intelligent investors except those hardy souls who deal in royalties for wealth far below the soil.

Can the grassland, once disturbed, be restored? If destroyed, can it return? Apparently so, but the greater the damage, as a rule, the longer the time and the greater the effort required. The underground parts of a great many of the native plants have considerable vitality, and may resume activity if given a chance, even after prolonged abuse. Seeds too will survive long waits for favorable conditions. The modern physician will tell you that in most cases the best he can do for his human patient is to clear the way and give nature a chance to make the cure. In the case of grasslands that have not been completely wiped out, such seems to be the best prescription for returning them to health and vigor. Protect them against the excessive grazing and fire that have brought about their ill health and let what remains of the original vegetation fight it out with weeds and other aliens. There is no reason to worry about the outcome. The rest cure will work marvels.

Weeds resemble those people who thrive best under difficulties and adversities. Prosperity and peace ruin them. They cannot retain their power under a calm and stable regime. Weeds, like wild-eyed anarchists, are the symptoms, not the real cause, of a disturbed order. When the Russian thistle swept down across the western ranges, the general opinion was that it was a devouring plague, crowding in and consuming the native plants. It was no such thing. The native vegetation had already been destroyed by the plow and thronging herds—the ground was vacated and the thistles took it over. It is the same with the American prickly pear, which is regarded as an unmitigated pest in Algeria and Australia. Again, the Mormons were accused of sowing sunflower seeds along their route as they crossed the continent. The truth is the sunflowers have always been there, but the hoofs and wheels of the Mormon trains broke through the turf, destroying it and giving the sunflowers their chance—a temporary one at best. No one ever saw a field, protected against fire, plow, and livestock, support a permanent population of thistles, sunflowers, or any other kind of weed. Just as surely as the weeds replaced the native plants with misuse, the reverse will happen under protected conditions.

Of course, this process of regeneration has its practical side. The land cannot continue to be tapped, any more than a cutover forest can be brought back to a commercial basis if the saplings are used for fence posts as fast as they appear. A longtime financial policy, looking more toward ultimate than immediate profits, must be established. Particularly will this be true if sheet erosion has removed necessary mineral nutrients from the soil. Usually phosphorus and even potassium and calcium have been lost and must be restored, and so must nitrogen. Again, gully erosion may have reached a point where engineering is required. These measures require expert counsel, labor, and generally an outlay of money. Whatever the initial cost of restoration, the recuperating acres have to be regarded as latent rather than productive wealth until again ready for rational use.

What may be done about those millions of acres, particularly in the drier grasslands, whose sod has been destroyed by the plow, but whose continued profitable cultivation is not possible? So long as there remains the most remote possibility that these can be made to yield crops under cultivation, we may count upon human stubbornness to return again and again to the attack unless there is some restraint. The slightest encouragement in the form of working capital will prolong the struggle, often to the final cost both of individuals and the commonwealth. If the best interests of all concerned depend upon getting these denuded lands back into grass, what chance is there to restore them?

Under favorable conditions it has been found that the planting of a grass known to be suited to local conditions and its subsequent careful protection will bring about a return of satisfactory pasture. Or if ample fields of buffalo grass are near at hand, narrow strips may be cut from them, separated into small blocks, and these transplanted at intervals onto the bare and abandoned space. With reasonable rainfall the native grass will establish itself and creep out to cover the clear intervals between the blocks. Both methods involve expense, waiting, and the hazard that drought may delay or prevent success.

But there is a great deal of loose, sandy land, once good pasture, that has blown and drifted until it has become a temporary desert. It is vital that some cover, no matter what, be developed here without delay. Nature has furnished a hint. Throughout this region after the drought was well begun the despised Russian thistle did so well that it was often the only plant available for stock feed. Instead of seeding the area with costly grass seed, whose success is a gamble, it might be sensible to mix in a good proportion of weed seeds. If the land is abandoned, weeds will be the first cover anyhow, and, as we have seen, they are transient settlers at most, preparing the way for the better kinds of plants. Actually the method is not new. There is a brilliant example of its use on the bare clay slopes of the huge Ohio Conservancy dams north of Dayton, which today are held in perfect condition by a dense, well-developed sod. The success of this plant cover was insured from the start by the deliberate use of the cheapest, weediest, mixtures of grass and clover seed that could be obtained. The weeds took hold at once, but their more genteel companions are now in full possession, just as the coonskin cap and leather jacket have been replaced by the fedora and business suit in the one-time wilderness beyond the Alleghenies.

Today in many places west of the 100th meridian much land is not only being restored, but converted to intensive use by irrigation from underground sources—a change that is strikingly obvious from the

air. Although a seeming miracle, it is only possible by
the use of fossil fuels in a time of growing energy shortage. Nor should a false optimism be allowed to conceal
the steady lowering of the water table where, as too
often, it is taking place.

9
From Longhorn to Combine

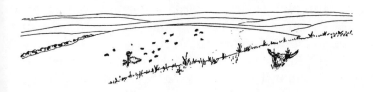

THE SHORTGRASS AND SAGEBRUSH country represent about the last stand of that legendary hero, the pioneer, unless one sees in the modern racketeer something more than is really there. Upon the cowpuncher fell the spiritual poncho of Adam Poe, Daniel Boone, and Kit Carson after these shadowy figures had passed to a still more shadowy beyond. From the days of the dime thriller to the "Western" his name has been bathed in glory—a source of constant inspiration and dependable profits. This being the case, exactly who were the cowboys?

Economically, they were hired hands, working for others. Personally, they ranged from the usual sort of hired hand one finds anywhere to charming and gifted exiles whose existence elsewhere had been fraught with disappointment. Frequently they had been the source rather than the victims of this disappointment. It was essential that they ride horseback and desirable that they develop sufficient skill and judgment to keep from getting their guts kicked or trampled out. Riding and even roping are more matters of practice than of divine inspiration. Beyond this the demigods of the range were kept in excellent physical condition by the kind of life they led. Their sense of proportion was kept in order by group discipline—a homemade sort of psychoanalysis known as "sandpapering the ego" or "applying the cactus." All told, they were a good lot, patrolling for their overlords a domain that reached west from the grain belt through the mountains, and extended from Canada to Mexico.

Their employers were enterprising capitalists, engaged in quantity production of beef and hides. By purchase or lease they enjoyed the grazing rights on vast areas. The grazing lands of a single owner might be together, but were frequently scattered, giving thus a better possibility of shifting the load with the seasonal and yearly changes in pasture and water supply. Of a necessity they were in close touch with the breeding and feeding of their animated wealth, and understood, whether or not they repected, the vital importance of conserving the range. The objects of resentment and envy from the land-hungry to the east of them, they labored under the added handicap of being unable to secure permanent control of much of the land that they used.

After 1870, with the development of transcontinental railroads and the systematic destruction of the buffalo, the livestock industry expanded with amazing rapidity through the plains area of the western states. There was no legitimate mechanism by which an individual owner could secure outright possession from the government of enough land to engage in this industry with any assurance of safety and profit.

The homesteads, while large compared to a typical

eastern farm, were too small for a semiarid region. They were also laid out in rectangles without regard to quality and available water—a folly that Major John Powell had warned against in 1876. The huge grants made to the great railroads were not consolidated, but scattered, consisting of alternating sections on opposite sides of the right-of-way. It was, of course, possible, but often difficult, for the wealthy operator to buy out a number of homesteaders and get the land he needed. If the land so purchased made a solid block, it was unusual.

The other recourse was to use the land of others, with or without rental. Gradually, therefore, a system of leased grazing rights developed throughout the entire western cattle country, resulting in serious, too often permanent, damage to the range. Here, as in the great lumbering operations of the same period, we see not a fair trial of a system of responsible private ownership, but the concession system operating in its worst form. Had it been possible to allot as permanent holdings sufficient areas to individual cattlemen, there is reason to think that the quality of the range might have been conserved. For the cattleman, unless he was a nonresident capitalist, differed in certain important respects from the lumber baron. The latter, for example, had about the same interest in trees that the butcher might be expected to have in cattle. The trees were ready at hand, but it had not been necessary to supervise their breeding, birth, and growth, or to move them about to prevent their dying of hunger and thirst. So far as the lumberman was concerned, nature could grow new trees while he made off to fresh sources of supply.

On the other hand, the chief responsibility of the

cattle industry was to grow the thing that was to be sold. This meant attention to sources of supply for food and water. The cattleman soon learned that a badly overloaded range meant inferior pasture in succeeding years. But since he so seldom had the title to his pasture, the temptation to make immediate profit while he could was too strong to resist. Resulting directly from this pattern of exploitation has been the periodic overexpansion of the cattle industry, more or less following on the intervals of heavier rainfall and more abundant grass. These events not only produced a surplus of cattle, ruining the market, but threw an abnormally heavy load on the range at the beginning of the several dry periods, furthering rapidly the ultimate destruction of the grassland turf. Under the system of land allotments, any planned and provident economy became well-nigh impossible.

It is also a matter of record that financial interests that had encouraged ranching west of the Missouri were, to put it mildly, far less zealous about seeing that settlers got a fair price for their cattle. In addition, the livestock industry has been faced with other and serious problems. Until recently, and as they will be in the near future, with the growing energy shortage, sheep have been as necessary to man as cattle, particularly where woolen clothing must be worn and where other sources of protein food need to be supplemented. Like goats, however, sheep are thin-lipped, sharp-biting animals capable of cropping very close to the ground. Their small hoofs are edged and will cut and wear the turf unless care is taken. On English golf courses where sheep have been used as lawn mowers, the moist climate encourages rapid growth of grass, and the animals keep it cropped to a close green velvet. If land in regions of high rainfall were not too valuable for sheep pasture, nor subject to damage from dogs, it would afford the best setting for that purpose. Actually, however, it generally becomes necessary to pasture great flocks of sheep in semiarid regions where the grasses have only a brief growing season, remaining the rest of the year dormant and dried—a sort of half-living hay—on the stalk. Under these conditions considerable care and expense is required to keep the flocks of sheep on the move and prevent lasting damage to the range. Moreover, cattle do not like to graze over ground that has been used by sheep, although recent experiments suggest that cattle will use the valley floor and sheep the adjacent high ground when both are in the same area.

At any rate, the entrance of sheep into the cattle country meant a declaration of war. A mild and kindly ranchman who, like Cowper's friend, would not needlessly set foot upon a worm, describes without batting an eye how great flocks of sheep, coming too near his territory, are inexplicably clubbed to death, or have their throats cut. This suggests again the very practical attitude of the cattleman towards conservation of the range.

Toward the would-be small homesteader he could scarcely defend himself by such direct action, although no doubt the intimate history of the West might produce instances in which he did so. Against the settler his only defense could be, in the long run, previous personal ownership of all of the land that he used. The weakness of his position lay in the fact that he was obliged to count on leasing large areas of the public domain to continue in business. And toward this domain hungry eyes were cast.

Probably the great transcontinental railroads, which at first provided a welcome outlet for his cattle, but which also fostered colonization, were inevitable sources of trouble. Mining operations often enriched him. At the same time, they helped build centers of population, creating customers, it is true, but also bringing in hordes of the land-hungry. In the end, it was pressure from this last group, stimulated by persons in search of economic and political gain, that caused the cattle industry to suffer most.

Yet in spite of difficulties the livestock industry made substantial strides toward permanence. The periodic maladjustments wiped out the plungers and exploiters. The large operators who have survived are those who have been sufficiently farsighted and conservative to prevent their being caught by cycles of overproduction, and in particular those who have steadily used their profits to acquire title to their own ranges. At their best, such men have developed a sense of responsibility for the welfare of the humans, the animals, and the land under them, comparing favorably to the best of the great plantation owners in the old South. An additional factor that contributed to stabilization of the industry was the recognition of priority lease rights to government land and, in particular, the enforcement of regulated grazing thereon. It is quite probable that if general conditions in the country at large could have remained economically balanced, the livestock industry might have shared in this general stability and the region that supported it might have developed an equilibrium. Such has not been the case. But before examining the latter end of the story, it is well to acquaint ourselves with a region whose history is as graphic as a diagram.

The Sand Hills of Nebraska contain about eleven million acres—an area larger than the agricultural portion of Egypt. These billowing, grass-covered hills lie in a vast rock bowl that holds the meager rainfall and slowly passes it up through the loose sand. In summer, when the pastures east of it are parched and dry, the grass here is green and fresh. But here and there, as fateful warnings that spoke plainly to the practiced eye, were great blowouts. These were funnel-shaped craters dug by the wind into the sand wherever the grass had been removed and the weak turf destroyed. Within these craters is a summertime inferno with temperatures often as high as 140° F., while even a moderate wind converts them into a withering, etching sandblast. With all her resources, Nature has a painful task to reclaim these blowouts. Given time, she can do it by means of the wiry, creeping rootstocks of Redfield's grass, followed slowly by other venturesome plants and ultimately by the original turf-forming grasses.

As long as this land remained in public domain, it was leased in large blocks and used as cattle range. Between the hills were numerous lakes where the underground water came to the surface, and about their shores were meadows of grass that could be cut and cured for winter feed. By the use of large areas for each individual operator, overloading of the pasture on the hilltops could have been prevented, as doubtless it frequently was, and the turf thus allowed to remain and hold the sand in place. But a kingdom like this, the size

of agricultural Egypt, was too tempting to be allowed
such use. It became a political issue; the Sand Hills
should be carved up into homesteads, each a mile
square, and given to the people. Finally, a man was
elected to Congress on the issue—Moses Kincaid. He
secured the passage of the necessary homestead law,
and the settlers who thronged in on these claims were
known after him as "Kincaiders".

The Kincaiders often lacked sufficient capital, as
well as previous experience with the kind of difficulties
that lay ahead of them. Most unfortunately, the area
assigned to each homestead—one square mile—was too
small to support a family under the conditions that
prevailed there. Some confined their activities to cattle, but were faced with the fact that enough cattle to
support them made too heavy a load on the range. The
close-cropped turf broke through, and the sand began
to blow, spreading ruin. Others boldly attempted to
plow the ground and plant crops. On the lowland there
was some return for this trouble, but at the expense of
the hay meadows. On the upland, the wind swept down
and across the planted rows, swirling the sand into the
leaves of the planted crop and shredding them to
pieces, finally either burying the crop or uncovering its
roots.

Easter week of 1920 a family of these homesteaders
extended their hospitality to two foot-travelers, with a
grace that would have adorned a more stately mansion
than the tiny sodhouse. In the house was a little flour,
a little coffee, a few potatoes, and happily, water with
which to prepare all three. No milk, no butter, no eggs,
no meat. Not one sign was there that the guests might
be making inroads on a scant provision—only regret

that the repast was not more varied. These brave people were the lords of 640 acres—a space that in Egypt keeps alive one thousand persons—yet they were slowly starving. When the travelers took their leave, they were cheerfully urged to return in August to help eat watermelons because "that's sure one thing we kin grow in these valleys, melons."

As the discouraged and defeated Kincaiders retreated from the picture, their homesteads were gradually acquired once more by the larger cattle operators, this time as owners instead of lessors. The pitted and scarred landscape gradually is resuming its proper function as a range region, but in the meantime the loss, both in potential wealth and in human effort and happiness, has been appalling. Here again, as in the monotonous story of exploitation we have rehearsed, it is not possible to fix individual blame. Certainly not on the homsteaders, or the cattlemen. Not on the well-meaning, humane politician whose name is associated with the experiment. Rather it rests upon a system that tolerates private privilege in utter disregard of public policy, and that which as yet does not demonstrate that science may be made to help in determining policy. At the time these measures were planned, there were men who knew the Sand Hills from the scientific side and who could have predicted exactly the outcome, but their views were not consulted in any effective way. Like the expert witnesses in our courts, scientists are often supposed to talk only when they have arguments for, not against, a popular or influential project.

The Sand Hills, although not covered with short-grass, but rather with bunch grasses, illustrate the situation in the true shortgrass country, which extends from about the 100th meridian west to the Rockies. In this country, of course, there are many fertile river bottoms where alfalfa and other crops can be grown. Such places have been enlarged by irrigation, at first by impoundments, more recently by the tapping of underground sources. Some of these sources are being recharged, but as distance from the great river beds that carry flood waters east from the Rocky Mountains increases, such irrigation is living on borrowed time.

As for most of the upland, its destiny is pasture providing range cattle to be finished for market either in feedlots or on farms where forage is grown and where plant, animal, and human industry can be combined. The cheapness with which extensive areas of this land could formerly be acquired was a temptation to speculators. "I saw the time," said an old banker, "when you could get a square mile of this for a pair of boots. The only trouble was, no one had the boots." Such were the conditions following a few years of the regularly recurring drought. On the other hand, given a few years of good rainfall, which comes as inevitably as the years of drought, things brighten up, and upland crops may be temporarily successful. At such times speculators could unload, and did so, in smaller parcels than were safe for the purchaser to live on. Let the dry years return, as inevitably they will, and the plowed soil moves on with the whirlwind.

These conditions were frightfully aggravated by the high prices for wheat during, and for a brief time after, World War I. From western Texas north through the Dakotas, large scale upland farming with the use of power machinery was hailed as "factories in the fields." Let it forever be recorded that there was ample information, based on the disastrous small-scale experiments that had been tried in western Kansas and Nebraska, to have forestalled the tragedy that occurred. But those were fabulous years, when the still, small voice of a president urging economy and forethought was put down as a bit of New Englandism that should be laughed off and not allowed to interfere seriously with the general popularity of his quaint, canny character.

During these years, the 1920's, there were for a time good wheat prices and favorable rainfall in conjunction. Everything in the country was going full blast. It was the most natural thing in the world for the plains farmers, whose cattle business had prospered during the war and who had been encouraged to try dry farming, to attempt the growing of wheat on a huge scale. The soil was loose and friable; the land was theirs to use as they saw fit; gang plows and discs could be dragged across the level with astounding speed and ease by huge tractors. The wheat grew well. Not more than knee-high, sometimes ankle-high, it could be cut by a header and threshed all in one operation, leaving the short stubble standing until again disked under. Although the yield per acre was not high compared to the eastern wheat land, the capital involved was less per acre, and the whole proceeding seemed very profitable. During thousands of years the slow growth of natural vegetation had stored considerable fertility in the soil. Amazing stories began to appear of the new era of power in agriculture—factory production applied to the land. As in the great spinning mills,

where only one attendant was required to keep numbers of machines in smooth-running order, here were
huge farms that could be operated with ease by
mechanical means.

For the small farmer of the eastern states, depression began in 1921, eight years before it struck the
urban economy. Many were driven out of business as
surely as those who own small factories or shops are
being "merged" or exterminated today. During those
years there was actually an astonishing decrease in
the number of acres of tilled land in the older states,
and the murmur of discontent about the farm depression, which had been growing since the close of the
war, swelled to a rumble that disturbed or delighted
the politicians, according to their particular commit-
ments. But out in the shortgrass country, make no
mistake about it, the wheat was being delivered.

Then, suddenly, everything collapsed. There was no
market for the wheat. Even before this disaster the
wheat farmers, although making money, were grumbl-
ing. They felt that the manufacturing and shipping
industries virtually had been handed gifts of money
from the government. In the words of one of them:
"Sure we are raising plenty of good wheat without a
great deal of expense, and we are getting along. But
these other fellows got rich, why can't we? If they can
have money handed out to them by the working of a
tariff and in other ways, why shouldn't we get all the
traffic will bear?" This authentic comment shows that
the "farm relief" business is one of many facets. Protest and distress are not always surely proportioned.
Among the wheat farmers there was growing resentment against the marketing agencies whose million-

bushel elevators levied a charge of one cent per bushel per month for storage but that, after all, were conducted with an eye to business and under conditions of severe competition. Cooperative storing and and marketing agencies were organized, many of them to fail because hard-working farmers of modest means did not see the importance of paying managers enough to secure the best and keep them honest. After the collapse of the wheat market much wheat was stored out-of-doors in great heaps, where weevils, disease, and weather were free to work.

The causes of the disaster were involved with the whole question of the depression. Competition from other continents, stagnation of the American foreign trade by high tariffs, and European impoverishment were, of course, important factors. But the one undoubted fact of overproduction in this country is what concerns us here. To an important extent this was accomplished by the wholesale exploitation of the shortgrass plains, whose permanent value for extensive grazing was a demonstrated fact and whose permanent value as a dry farming center remained to be proved.

Artificial restoratives were applied to relieve the situation. These helped steady matters somewhat, but Mother Nature had a remedy of her own, one that shortly appeared. Climate, which had been passing through the humid phase of its cycle, swung back towards the dry side. Months passed without a drop of rain in western Kansas in the summer of 1933. The next year was even worse. Not only did the wheat fail to mature, but so little grass was left unplowed that the livestock was starving and had to be moved or killed. Buyers with a little cash could go through this country and name their own prices, sheep two bits apiece, cattle five to ten dollars. From the first of the drought there was considerable wind sculpture of the naked fields and local dust storms such as one finds in a dune area. But by the spring of 1935, as the dry cycle continued, these dust storms became matters of national extent, fitly symbolizing to distant Washington and New York the painful distress that existed in the country the dust came from. Whether the dust that fell in a given spot in the East actually came from the plowed-up pastures of the West, or from the desert itself, farther west, is immaterial. Within the newest wheat region fences hidden in dust, houses buried to the eaves, blinded jack rabbits, and suffering humanity were tragic proof of the penalty incurred by disturbing nature's hard-won balance.

The drought, which was the apparent cause of the disaster, was certainly predictable—not in any exact sense, of course, but as unavoidably due to occur at intervals. A system of agriculture had been put into operation in disregard of the certain hazards of the shortgrass region, and the dust storms became the costly, spectacular evidence of this fact.

10

Dust

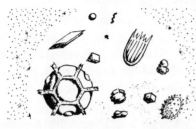

On a day in March, 1935, the meticulous Ohio housewife, whose weekly dusting had always been sufficient to keep things trim between housecleanings, suddenly finds herself unable to deal with the ubiquitous, curiously ruddy dust that is settling down out of the upper air and filtering through her tightly built establishment. She knows that dust is one of the normal tribulations of her sisters who live in the West. She has heard of the old physician in southwestern Nebraska whose diary, after his death, was found to contain the following entry: "Wind forty miles an hour and hot as Hell. Two Kansas farms go by every minute." Now her radio tells her of great dust clouds in New York and Washington. The stories in the old school geography about the sandstorms of the desert had always seemed somewhat too fantastic, describing as they did the camels with their funny trapdoor nostrils and the Arab with his scarf ever ready to serve as a dust mask. Yet here is today's paper, saying that out on the Great Plains, where that red dust on her library table comes from, a child has been found stifled in a windrow of dust. And wirephotos show schoolchildren with faces veiled like little Fatimas as they trudge along the country roads out West. Pulmonary troubles, especially pneumonia, are reported on the increase because of the dust. One little town, where funerals are rare events, has four in one week, and hospitals near the stricken area are having plenty to do. People are reported to be moving out with their household goods on trucks. Each day's conditions are chronicled in terms of visibility—one block, a quarter-mile, or, on better days, perhaps a mile.

As the dust confuses the senses, so the welter of reports and opinion clouds the judgment. It is a reporter's paradise, an editor's purgatory. The owner of a large tract in the shortgrass country, which had been plowed up for wheat, says:

> We're through. It's worse than the papers say. Our fences are buried, the house hidden to the eaves, and our pasture which was kept from blowing by the grass, has been buried and is worthless now. We see what a mistake it was to plow up all that land, but it's too late to do anything about it.

Others, who know the country well, say that these storms are nothing new. They have happened before. If the two-year drought could be broken, say they, the grass would start up again and all would be well. The area is on its way toward becoming a vast cultural desert, like parts of western China, says one recognized expert. Others, also qualified to know, say that the actual damage is comparatively slight, the soil being mostly so rich and deep that the loss of the top means little, except on certain soil types.

A former president, traveling through the stricken region by train, is quoted as saying, "Too much of the grassland has been plowed up." And an able agronomist says the trouble is due not to plowing, but the lack of it. He blames the shallow disking, and recommends plowing the land into deep furrows, provided that new ones be made as fast as the wind levels up the old. An enterprising reporter recalls the wave of indignation against a cabinet member who had once referred to the wind-ridden area as "submarginal" and who proposed moving the inhabitants elsewhere. Editors incline to play down the general serious character of the whole matter, yet not enough to discourage a liberal-handed government from coming to the rescue. A few people may be moving out of the stricken region, they admit, but the sturdy old pioneer spirit will not down—we shall see the dusty West once more the granary of the nation, provided a little timely help is forthcoming. It is hard to tell which is the real show—the procession of news on the front page or the kaleidoscope of opinion elsewhere in the paper.

As a matter of fact, so huge is the region within which the dust is being picked up that any particular statement may be at once true and false. Let news staffs and advertising managers wrangle, while chambers of commerce and eminent economists disagree. What if the stricken farmer has a different verdict from that of the impersonal scientist? What if one township, or one thousand, will never grow wheat again? The truth in any particular sense is only important as it fits into the larger picture. The really practical thing is not the immediately practical. To regard the dust plague as a misfortune brought on by the wrong kind of plow, or a chance drought, or even to concede that it is caused by an unhappy shift from pasture to wheat, is to miss the point.

Unless the dust is seen as a symptom and a symbol, instead of a direct problem in itself, the misery that it has caused is of no avail. It is precisely this clever, efficient, and speedy solving of immediate problems, without regard to their general setting, that has brought us where we are. We have stretched to its limits the merely opportune. Let us not forget that the last bumper crop of wheat from the dry farming country lay beside the railroad tracks in vast heaps for months while people willing to work for food were not getting the chance to do it and were hungry. If tomorrow some ingenious fellow were to insure that the soil of the dry West would never again be lifted into the air, it would be of no solution to the real problem that lies back of the dust storms of this year of grace, 1935.

Dust itself is nothing new. Like the circle, it is a symbol of eternal time. Long before the days of the microscope and the chemical balance, it was understood that dust is the beginning and the end of all things. Dust is always in the air we breathe, an invisible world of tiny, buoyant particles, infinitely rich in its variety, and with laws of its own. While most people think of it as being only minute bits of earth stirred up by strong air currents, it contains a host of living organisms, bacteria, molds, pollen, animals, as well as fragments of material from larger plants and animals. Except perhaps in air newly washed by rain, these particles float about perpetually sustained by gentle drifts in the atmosphere of which the human senses are scarcely aware. Even such a giant citizen of the world of dust as the plumed dandelion fruit can remain afloat indefinitely in a breeze of not more than three miles an hour. The microscopic grains of pollen and fungus spores, capable of ascent to the stratosphere, have their own curious globular symmetry, often richly marked, but with neither right nor left, top nor bottom, as fitted to the ocean of air as the fish, the fly, the elephant may be to the respective worlds in which they live. With the rising and falling of the currents, the particles of dust, living and dead, are perpetually renewed, perpetually settling out. The world of dust is never at rest, and the potential energy of the surfaces of the granules in it is enormous. One of the serious hazards of the grain elevator business is the peril of spontaneous dust explosion from sparks of static electricity within the dust itself. (Forty-three years after the first publication of this sentence a series of disastrous grain elevator explosions took place, despite the existence of an inspection service and knowledge of the economic hazard involved. This is a prime example of the difficulty of reconciling effective controls, private enterprise, and scientific technology for longtime benefit, in a nation as great and diverse as the United States.) The behavior of dust is influenced by electrical conditions to a surprising extent. Dust poisons used to kill plant pests will not stick in place unless they bear an electrical charge of opposite sign to that of the leaf surface.

At times the dust world receives important accessions. This often happens as the result of volcanic explosions. When the great volcano on Krakatau blew up in 1883 it not only buried the entire island in a deep mantle of ash but sent clouds of it into the upper air where they hung for months, girdling the earth before they slowly settled. The eruption of Mont Pelee on the Island of Martinique in 1902 was followed by a season of voluptuous sunsets occasioned by the impalpable dust that filled the atmosphere above North America, spreading the dying rays of light each evening into vast masses of gold and purple. Likewise, in 1912, the dust from the Alaskan volcano Katmai spread far and high, although the heavier bulk of it dropped within a radius of one hundred miles, enfolding every living thing beneath a thick blanket, and apparently creating a complete desert. Presently, however, insects appeared in countless numbers, and later it was found that a good deal of plant life had managed to survive in one form or another to emerge and slowly reclothe the seeming waste.

In each of the three great eruptions mentioned, the succeeding year was marked by abnormally low temperatures that have been attributed to the interference of the atmospheric dust with the radiation of the sun.

While the movement of dust by the wind is usually associated with dry, treeless climates, there may be very violent activity of this sort in moister, more genial conditions. Along the shores of the Great Lakes, particularly Michigan, enormous dunes of sand were so built up. Eventually, most of them were stabilized by a forest cover, but so feeble is the hold that this barrier maintains that the slightest disturbance of the ground cover has been sufficient to start forming the huge blowouts. When the city of Gary, Indiana, was created upon the dunes, the blowing sand persisted in marching through the town. Had it not been disposed of, it might have swallowed the streets just as it has the forests of pine and oak nearby. When this carving of the sand is under way, it forms a blast that few plants can withstand, shredding leaves and stripping off the bark, leaving trunks bare and polished on the windward side. Only two kinds of plants can resist: those that, like certain vines, can drop as the sand is scooped away below them and form fresh roots continually and those that can grow fast enough at the top to survive burial, meanwhile forming new roots just below the rising ground line. Because the sand grains are comparatively heavy, they are seldom carried far in a single flight, although they move many miles in the course of time by repeated short jumps.

One also sees fierce, though local, dust storms on the cultivated, powder-black soil of old lakes and peat bogs in the northeastern United States. Mostly these are used, after drainage, for such clean-cultivated truck crops as mint, onions, or celery, and in the course of time the muck becomes very finely disintegrated. The so-called muck-blows that occur in high winds are not only exceedingly unpleasant, but may be disastrous to valuable crops, particularly when the plants are young and tender. They also tend to injure the field itself by removal of the soil that gives it its peculiar value.

Plowed fields of any kind that are not covered during a time of high winds are apt to be blown about, even in a normally moist climate, especially if they contain much sand and silt in proportion to clay and humus. Roads, too, which in the aggregate cover a considerable area, contribute a substantial share of dust. Military men know that dust often betrays the position of marching columns at long distances, while the traditional kerchief of the cowboy had its origin as a dust mask, used on the trail behind the milling cattle.

A few decades ago, when macadamized roads of crushed limestone were being built everywhere, the white dust from them was often very troublesome in the summer months. Before many years white sweet clover, until that time an inconspicuous weed, literally swept over the countryside, particularly along the shoulders and roadsides of the macadamized "pikes." Bees thronged the heavy-scented blossoms, and pampered family horses would loaf along, greedily snatching the nutritious tops. The clover appeared because of the lime that the road dust had added to the adjacent soil, a substance necessary to any legume if it is to thrive. The clover is gone now, replaced by more stable vegetation, but the soil is better because of its sojourn. This episode had the double benefit of attracting attention to the value of sweet clover as a crop and of emphasizing the shortage of lime in many agricultural soils of the eastern states. Similarly in Texas, exhausted and abandoned rice fields seem to recover most rapidly within the dust zone of the shell roads that pass through them.

We have said that dust is nothing new. Those who know the banks of the Missouri at Council Bluffs, or who have observed the Mississippi above Natchez, are familiar with the vertical, pale brown cliffs of material almost as fine as talcum powder. This material, known as *loess,* is found at many places in the Middle West, and was dropped where it is by the wind. There are some differences of opinion as to where it came from, some holding that it was the fine flour from the glacial mill, washed out, dried, and shifted by the wind, but not carried very far from the place of origin. Others think it came from far to the west. Evidently, it lodged where it is because there were plants whose tops caught and held it. One can see this same sort of slow building up at many places in the grasslands today. At any rate, it is a good example of the geological importance of wind as a mover of dust.

Whatever theories may be held regarding the source of this loess, there is little doubt that it was moved during very dry periods of climate, probably while the great glaciers were retreating. During such dry periods there is the least possible interference from vegetation with the lifting of soil, for deserts are at their maximum size. It is quite possible that during these very dry climatic periods the country west of the Mississippi and Missouri valleys was much barer than today, and that the westernmost vegetation heavy enough to lodge and arrest the blowing loess was located approximately where we see its deposits now. At the foot of the Medicine Bow Mountains in southeastern Wyoming, there is today a huge basin some 10 miles long, 3 miles wide, and 150 feet deep that was certainly scooped out by the wind. Its former contents are now lying in Kansas and Nebraska, whether or not they are mingled with fine material from the melted glaciers. Elsewhere along the Rocky Mountains within the Great Basin and the loess plains of Northeastern China, there is evidence enough to prove to the most skeptical how long and hard the winds have worked.

That the winds are still severe and constant enough in the interior of the continent to move an immense amount of material, no one who has been there will question. Where they have long had a free sweep at the naked surface of the earth, not much is left except coarser material that will not travel far. This is the explanation of the stony surfaces known as desert pavements. But wherever the interior of the continent has been clothed with turf, the situation is vastly different. Here the surface material is characteristically fine, and has continually become more so through the physical and chemical processes of soil formation going on among the roots. Yet these same roots, like straw in adobe bricks, have held the fine material together against the steady sweep of untrammeled winds. The margin of control is not a large one. In the more critical areas the roots seldom get deeper than twelve or eighteen inches and the vegetation itself is none too dense. There is no saturated water table close below to entice the roots deeper and bind the soil by wetting it. The deeper one goes the drier it gets. Such is the thin, brave line of roots that holds the outposts of our productive land despite all the vicissitudes of alternating drought and favorable seasons. Properly husbanded, it is capable of yielding meanwhile a return, modest by the acre, but bountiful in the aggregate, in the form of grazing. It is no piece of maudlin sentiment to say that this frontier of vegetation deserves well of humanity.

Had we accorded it fair treatment, not even the prolonged drought of two years would have released its soil into choking, impenetrable walls, sweeping over the continent and far out to sea. The reckless overgrazing that began in 1870 and has continued since, with periodic overproduction of livestock and steady destruction of the native turf, was bad enough in all conscience. But when, from Dakota to Texas, the multiplied power and mechanical perfection of modern engineering was loosed, in competition against already struggling farmers in more favorable climates, to destroy the sod and replace it with wheat, the outcome was inevitable. This outcome was not only predictable by anyone who knew the vegetation and climate of North America, but was predicted without causing anything but resentment. With the turf gone and the cycle of moisture past its peak, with the winds maintaining their normal behavior, the country literally started to blow out of the ground. For this great catastrophe the individuals directly responsible have paid in bitter coin, and we all shall have to pay in a measure. No work of ignorance or malice is this, but the inevitable result of a system that has ever encouraged immediate efficiency without regard to ultimate consequences.

Heaven knows that flood can be frightful enough. But, after all, there is something tangible about water. It may wash away your property, or drown your friends, but you can see it and know how it works. You know whence it comes and whither it goes. The dust is different, today from New Mexico, tomorrow from Wyoming. It blackens the air and hides the sun. It does not bring the sharp, quick, desperate terror of flood, but instead a slow, chilling, and pervasive horror, perhaps out of keeping with any immediate damage, but right enough in the long run.

Mile-high, these gloomy curtains of dust were the proper backdrops for the tragedy that was on the boards. The lustful march of the white race across the virgin continent, strewn with ruined forests, polluted streams, gullied fields, stained by the breaking of treaties and titanic greed, can no longer be disguised behind the camouflage that we call civilization. Yet to say this, is not to be blind to the beautiful meaning, clear to the discerning eye even in the most impoverished homestead or most sordid, hopeless-seeming village of the New World. Sword and spirit never march far apart—priest and pirate, renegade and missionary—the best and the worst are in the vanguard as mankind moves along. And the general sort who move behind them are warped by the curse of the one, even as they are blessed with the light of the other.

11

Mud

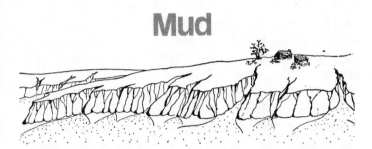

It is an early morning in March, 1913—a Monday morning to be exact. About three days of the usual spring rains had followed the melting of heavy snows, but the air is clear, and students in the quiet Ohio college town start as usual toward their first-hour classes. But there are no classes. The streets around the gentle knoll on which the college stands are a sea of swirling, swishing liquid mud.

For three blocks on either side of the usually modest Olentangy River a flood is raging. Amazing, preposterous, unheard of, a grand show until two men, clinging to a telegraph pole, are swept to their death, and other lives are being lost. Families marooned in upper rooms or on roofs are shouting across the water for help. Boats put out to rescue them are swamped at once by the powerful current. A house, split vertically in two with its halves still upright, floats south in midstream beyond hope of rescue.

Something has to be done; everything possible is tried, but there is no precedent for effective action.

Finally, a future major general shows his talent for command and organization, restoring order. Then an elderly, semi-invalid and former oarsman suggests that if a skiff were trucked upstream and launched, he could guide it to beleaguered houses, take aboard their inhabitants, and guide his craft to some safe landing below the town. This shuttle continues throughout the day; happily, a none-too-sturdy heart withstands the strain.

The flood subsides. Patrols working up and down the valley, feverishly at first (they found a living baby strapped in its crib that way) and later methodically, report that rich farms have been stripped to their clay skeletons in some places, or buried in a desolation of gravel in others. Within the town, cellars, streets, and near-the-river living quarters are coated with stinking slime brought from upstream in quanitities no one ever thinks to calculate.

An energetic governor inaugurates engineering measures that are soundly planned to minimize destruction whenever, if ever, such a flood occurs again. The "tree cranks" of the state point out that the flood was much more severe than it would have been if more of the forests were still in place. The pious speak—according to their conscience and their bile—of a "visitation of Providence" or the "wrath of the Almighty."

With variations more or less dramatic and sometimes tragic, the story repeats itself in one place after another each year throughout the length and breadth of the land. Often, as in Ohio in 1913, it strikes without warning and finds the populace unready. Generally, however, the annual flood is taken more or less as a matter of course. The people in the lower Mississippi Valley, like villagers at the base of a volcano, become habituated to the risk they run. It is a startling experience for the visitor to New Orleans when he learns that the great river is flowing along beside the city and above it, restrained by giant levees. And when this valley in the air becomes almost brimful, it is hard to understand how his urbane hosts sleep as well at night as they do. Of course, the Army Engineers are standing guard. And a great spillway has lately been constructed to divert the highest floods away from the main channel, out into the swamps that fringe the Gulf of Mexico. It may no longer be necessary to dynamite the levee far upstream, flooding out farms and villages to save the Crescent City. But the river remains a mighty thing, heedless of man.

The spectacular thing about flood is the rush of water and the threat of disaster to the lower valleys. It is natural enough then that the first efforts go toward preventing such disaster by taking measures locally in the places most apt to suffer. The problem presents itself there to the engineer in concrete form, within a fairly small space. His energies are directed not to preventing the coming of the water, but to handling it when it does come. If he followed it upstream at floodtide, he would find his efforts diffusing out through all sorts of tributaries and eventually onto the ranches, fields, farms, and forests of the interior. Actually, of course, the flood problem begins at these sources. There is no ultimate control unless it is there. All else is palliation. But the engineer is not to be censured if he prefers to avoid all the complications that would be his if he tracked the trouble to its point of origin.

Above the limits of navigation, authority is divided. Prudent action is complicated, even thwarted, by powerful industrial and political combinations more interested in immediate action than in Morgan's "conclusive engineering analysis." Membership in the Rivers and Harbors Congress includes contractors, suppliers, and many members of Congress. Local support, easily obtained, does not always measure merit; if this were not so, the expression "pork barrel" would be merely an insult.

If the safety of property owners in the lower valley only were to be considered, it might be fair to ask how far anyone in the upper watershed should be obliged to alter his management in the interest of flood prevention. Hope for improvement lies in the fact that these owners with whom the trouble starts should be just as much concerned as anyone. For the floods that threaten the lives and property of their fellow citizens downstream are steadily robbing the upper reaches of wealth in the form of water and soil, neither of which can be easily spared.

To find limpid streams today, one must go to the mountains or to sources fed by springs. In pioneer days the larger rivers were often clear and clean. Today they are opaque, even in quiet times of low water. At flood time, if one lowers a bucket into the Canadian, for example, and then brings it up and allows it to settle, as much as one-fourth of it may prove to be mud. If the property owner upstream could be made to realize that this mud is the same stuff for which he paid by the acre in cold cash, the flood problem would be many steps nearer solution than it is today!

The work of water in leveling the hills and carrying

them to the ocean bed has gone on since the beginning of geological time and must continue to the end of it. It is a normal enough process, and certainly inevitable; "flood plain" means exactly what it says. The danger of the present situation is not that any new force is at work, but rather that the rate of its action has been speeded up far beyond the capacity of nature to replace the eroded soil. Consider the framework in which events are taking place. Most of the continent is more than a thousand feet above sea level and in the Mississippi Valley, a great deal of it much higher. In the interior the rainfall is much less than on the coast and the vegetation in consequence much sparser, affording less natural protection to the surface. Such rainfall as comes on this meager cover is apt to come infrequently, but in torrents. Moving northwest through Oklahoma, for example, from one corner to the opposite, one rises from about five hundred feet to nearly five thousand feet above sea level in the space of some five hundred miles, an average rise of nearly ten feet to a mile. The rainfall drops from forty inches to about fifteen inches a year. The vegetation changes from dense forest with gum and cypress swamps to shortgrass and scrub. The conditions for heavy erosion are perfect. The infrequent but torrential rains at the northwest exert the maximum force, yet encounter the least resistance from vegetation. The slope is relatively steep, increasing the opportunities for erosion as the water proceeds on southeast through the state. Let man enter this picture and recklessly destroy much of the protective ground cover, and no prophet is needed to see what a tremendous impulse he will give to an already rapid process.

Erosion, like many another curse of humanity, grows by what it feeds upon. It behaves like compound interest. Beginning first on the exposed surface of normal soil, it first removes the spongelike, water-holding layer of dark humus, which normally is held in place by the roots of plants and protected by their tops. Once this humus is washed away, nothing remains to absorb the water that falls thereafter. Driving along the concrete highway after a heavy rain, one sees pale patches of ground here and there that dry off almost as rapidly as the slab of road itself, while the rest of the fields remain dark and wet. These dry, light-colored, almost waterproof layers are the so-called B-horizon of the soil, from which the humus-bearing, absorptive layer has been scalped by erosion. It is problem enough to have the highways, aggregating well over 1 percent of the contiguous states, shedding water like a duck's back into our drainage systems. But when we add the areas from which soil has been removed, which in consequence are unable to retain water, we have added enormously to the destructive power of our abnormally swollen streams. Thus does erosion promote still more erosion.

What is true of this scalping of soil—sheet erosion— is likewise true of the more familiar and spectacular type—gully erosion. Both grow by what they feed upon. As a gully cuts back, tributary gullies are cut and the damage spreads like a ringworm, in a circle. The measure of damage is not the *distance* cut back by the main gully, but the increasing *area* involved in the whole system. The more it eats the more it wants. The injury then increases, not by addition, but by multiplication—a truly frightful thing to contemplate. The end of gullying comes when badlands are formed, grotesque spires and chimneys of rock or soil standing like pegs on a board, with no stable spot where plants can grow and soil form to arrest the procession of ruin.

Gullies and badlands are the final effective protest of outraged nature, compelling the abandonment of land and speaking their story to the dullest. But in truth they seldom appear until after the mischief has been wrought that converts fertile soil into waste. It is the quiet, generally unnoticed sheet erosion that is the true measure of deterioration over much of the United States. In the early days of the last century, and in fact until recently, the prevalent notion of soil fertility was based upon its chemical character. Without doubt, the necessary mineral elements must be present, and can be readily exhausted by unwise agricultural management. If the same crop is grown repeatedly in a field, or if the phosphorus, nitrogen, and potash are consistently sold off the farm, whether on the hoof or in bushel baskets, the soil will be depleted unless minerals are added. But in actual practice it has quite generally happened, especially on rolling land, that more of the valuable minerals were washed away along with the surface soil in the runoff water than were ever sold off the land in the form of crops or livestock. Yet so subtly does sheet erosion do its deadly work that the steadily falling yields that have moved the frontier of abandoned farmlands westward have been generally attributed to loss of fertility caused by continued cropping alone. Just as a tree may have a rotten heart, but still look good from the outside, so may the surface of a field be unfurrowed by gullies, fair and even to look upon, yet worthless to the farmer because the best part of it has been washed off.

Equally as insidious as this quiet removal of upland soil is the fate of the lowlands when erosion starts. The level upper terraces of our old stream valleys have always furnished the choicest of farmlands because of the abundant moisture below, the protection afforded by the valley sides, and the rich deep soil. This black soil is partly due to the greater growth of natural vegetation, partly to the gentle movement downward from the hills of much good material. But when sheet erosion begins on the surrounding upland fields, this downward movement is no longer gentle. At first it actually enriches the valley, by bringing down the topsoil from above. Soon, however, this phase passes and infertile subsoil washes down, forming a pale coat over the dark valley land. Before long the infertile coat becomes so thick the plow cannot reach the dark soil buried below it, and there is an immediate decrease in the yield. This process, so simple to describe and so easy to see when one is looking for it, happens in such gradual stages that the farmer himself does not really understand what is going on. He only knows that the farm has played out, top and bottom, and if he is able to do so, he gets away from it.

Great is the power of water, either for good or ill. The tumbling mountain freshets in late spring knock huge boulders about with the ease of a boy shaking a pocketful of marbles. The same stream, harnessed and regulated, will drive turbine engines with precision and speed. In placer mining, the concentrated force of a small jet will strip away the side of a mountain. The same water, spreading out gently through ditches across the barren plains at the foot of the mountains, will convert them into rich green farms. Water is indispensable. What hope is there to control it in the interests of civilization?

This is not a problem solely for the engineer. It is as much a matter of biology as it is of engineering. It will never be solved until the engineer and the student of plant life attack it together. The delicate, threadlike roots of plants are as vital to a solution as are mighty walls of reinforced concrete, and the black and spongy soil must play a part no less than vast reservoirs. In fact, as we have seen, the reservoirs themselves soon become useless unless the valleys that drain into them are carpeted with plant life to retard the inwash of silt. Of course any inland pond or lake must fill up in the course of time. A foot of fill in three hundred years is not unusual under undisturbed conditions. Yet behind nearly a score of dams in the Piedmont country there has been a foot of silt added *each* year, until the reservoirs themselves have been rendered worthless in less than thirty years.

Can we check the erosion now going on? And what hope have we of restoring the enormous area at present rendered worthless by this process? Prevention, if started in time, is a cheap expedient compared with the cost of redemption after the mischief has been done. To prepare farmland when first cultivated so that it will not suffer from erosion costs far less than the amount needed for engineering, fertilization, and reconstruction of average eroded land not beyond repair.

Farm practice, whether good or bad, is usually the

outcome of apprenticeship that begins in childhood. Never easy, once formed, it is not lightly changed. This makes the farmer vulnerable to his own errors as well as to the vagaries of nature and society. Dealing as he does with growing plants and animals, he must look ahead further than the average small businessman who is often obliged to turn over his inventory three or four times a year to remain solvent.

Both farmer and retailer must work with a small margin of profit. Neither is in a position to do much about the long view even if he understands its importance. The difficulty increases where society gives top priority to immediate profit.

There are, however, communities where land is regarded as a trust rather than as a chattel to be treated as its owner sees fit, regardless of consequences. Here the atmosphere discourages sacrifice of future benefit to short-term advantage—a situation evident to the traveler in Scandinavia and the Netherlands as well as in ethnic communities in a number of parts of the United States. In such neighborhoods stewardship of the land has all the force of a religious principle, reenforcing rather than discouraging sound land use by the individual.

Without air we die at once; given air but no water we live for a time; without food but with air and water we may survive some days. But food (and fiber) from the soil we must have. Although, as John Taylor (1753–1824) reminded our fathers, the well-being of the farmer and preservation of the soil are essential to the well-being of the nation, it is a curious fact that there was no federal erosion-control agency until 1933, thirty-five years after the precursor of the Forest Service was set up.

Colonial agriculture brought to America from Britain and Spain was primitive and inefficient, with little skill or concern for proper care of the soil. Not until the beginning of the eighteenth century did better husbandry—long practiced on the Continent—become a suitable and fashionable topic for polite and educated society in Britain. From there it spread to the New World where farms had already been "worn out." Statesmen and naturalists began to speak out against soil depletion and loss, fortunate if they met only indifference rather than ridicule, for the supply of new land seemed endless.

We know today that there is no one sovereign remedy, but that a combination of methods must be employed, methods that may be skillfully adjusted to the particular situation. Many of the measures are of such a character as to be highly desirable for reasons other than the immediate control of erosion itself, so that they may form part of an integrated policy of effective land utilization.

On the engineering side these measures involve the erection of barriers to check the growth of gullies where they now exist, together with the construction of series of reservoirs of all sizes at convenient places to impound the excess water that today swells the rivers in times of flood. Fields that are not level must be protected by terraces arranged in such a way that the water from them is not dumped out onto the public highways, but retained on cultivated ground. The highways themselves must be scientifically planned so that they do not act as stream beds in times of heavy rain. Water must drain into broad, shallow, grassed roadside ditches and, where possible, into adjoining fields where it is needed. The technique of supplemental irrigation at low cost to farms in the moister, better farm areas of the country, where irrigation is today not thought of, must be developed. These are some of the problems the engineer must face. As to flood plains, the ancient prerogatives of their rivers should be respected. Building, especially of housing developments, should be forbidden and uses that can be productive between times of high water encouraged.

To perfect these measures, the specialist on soils and plant life must cooperate. His task is in a sense more difficult, for it speedily involves the delicate political problem of specifying the way in which land must be utilized. He runs squarely into the ingrained American tradition that land is private property with which the owner is free to do as he pleases. Before stirring up that particular hornet's nest, let us see what should be done rather than how it may be accomplished.

The first thing the natural scientist is likely to recommend, and the last thing as well, can be simply stated. Get vegetation back on the ground. Mother Earth is a staid and dignified old lady, no nudist by choice. Rough and broken land at the sources of streams and tributaries should be permitted to revert to forest and grassland without struggling against the handicap of plow, fire, or reckless grazing. This is only common sense, anyhow, considering our need for more timber and pasture than we have today. Increase the proportion of pasture on each farm, in particular on the rough, uneven ground that is hazardous to farm. Confine the clean-cultivated row crops as far as possible to rich, level soils not likely to be washed or blown away. Where rolling land is farmed, do not rely upon terraces alone, nor indulge in the vanity of straight furrows. Plow with the contour of the land, and on the terrace ridges use close-planted crops that form a tight, protective sod, alternating with more open crops on the level terraces themselves. When pastures are established, nurse them with the care a corporation puts into its manufacturing plant, or a banker into his (personal) investments. Gauge the size of flocks and herds, not by what the traffic will bear in good years, but on the basis of an average, wholesome, sustained yield. Practice wise rotation and diversity of crops, and see that in place of the minerals removed from the soil, a return in kind is made. These, which are, after all, strangely like the counsels of Hesiod the Greek and Cato the Roman, are the promptings of the modern man of science. Will they too go unheeded?

12

Wet Deserts

THE TRUE MEASURE of the desert is not in its water nor its sand but in the kind and pattern of its life. For deserts are rarely lifeless, however sparse and bizarre that life may be. Desert plants of the Old and New Worlds, seldom closely related, are much alike in structure. Pulpy shoots with coats of heavy wax spread their roots near the surface to catch the rare and precious rainfall, storing water against inevitable drought and spacing themselves to share it. In fact, the mark of desert is the presence of bare ground between plants. Spines as well as nauseous tastes and odors serve to discourage hungry browsers.

Like plants, animals of the desert are remarkably fitted to rugged conditions. Some never meet with free water. Most of them generally use shade and underground shelters to protect themselves from the intense heat and evaporating power of the sun. A rattlesnake can die from brief exposure at midday. Shreds of dry plant material absorb the moisture that condenses during cool nights, making it available for insects and other animals. And—astronauts note—desert animals may condense and reuse the moisture released in respiration.

Seeds of annual flowering plants remain dormant for many years, arousing and growing into glorious profusion whenever sufficient soil moisture permits. Until the supply is exhausted, these plants continue to grow, flower, and form seed ready to survive another long quiescence.

As to the sand so often associated with ideas of desert, inevitable setting of the sheiks of filmdom, Lawrence of Arabia tells us plainly of the stretches of rock and clay over which he fought. And as to water, the desert may have in places or at times considerable amounts. But this water may come too seldom or at the wrong time; it may evaporate or run off too fast, or it may pick up and dissolve chemicals unsuited for the use of living things.

In the Southwest of North America and the interior of Asia are desert basins that held lakes during the moist climates that ended with the retreat of enormous continental glaciers some ten millennia ago. As rainfall decreased thereafter, these lakes began to dry up, leaving highly concentrated solutions of the salts they had contained. Plants that survive in these saline or alkali basins reveal by their structure that water, though present, is hard to come by; they may be puffed up with hoarded water, gray with wax or wool to keep it from being lost, and widely spaced as are plants in waterless deserts.

Even where conditions are too severe for larger plants, there are other forms of life, visible only as masses of tiny individuals. Some grow in hot springs not much below the boiling point. Others may live as thin green layers in desert salt.

So much for deserts as man finds them under natural conditions. But our concern is with deserts that have resulted from human activity, and at this point, those that have no lack of water. For we have been no less heedless with the teeming wealth of life in lake and stream, pond and bay, than with that of forest and plain. To the pioneer, as to the Indian whom he so largely displaced, food from inland and coastal waters was so abundant that no thought was given to the possibility that it might not last forever. Neighbors would own a seine in common and in the autumn would join to harvest fish for winter use. These would be thrown into piles as nearly equal as might be, one heap for each family. By lot or blindfold choice each heap was parceled out and taken home to be salted or smoked. Spring-fed brooks were alive with fish, while larger streams and lakes contained seemingly inexhaustible supplies.

Today river seining is unlawful, and even the most skillful use of hook and line is too often ill-rewarded. Fishing, one of the most popular of recreations, is kept going by huge investment in hatching and restocking inland waters. This enterprise is financed mainly by income from those diplomas of vain hope known as fishing licenses whose sales, doubled between 1950 and 1970, have been increasing by a million each year thereafter. The theory has been that scarcity was the result of overfishing, but the truth is that while the fisherman has not been guiltless, he is not the real culprit. The number of reported fish kills caused by polluted water increased fivefold in the five years beginning in 1970.

Life within the water is a delicately balanced affair. This can be shown by sealing a live fish, a snail, and a water plant within a globe of water, placed where the sun can strike it each day. Such a little world will flourish for months, each inhabitant playing his appropriate part in maintaining the proper physical and chemical balance. The plant manufactures food from what is, to the snail and fish, waste material, and these in turn consume the food. While making the food, the plant releases oxygen, thus purifying the water so that the animals will not suffocate. Some of the waste from the fish is not in a proper form for the plant to use directly, but is made so by the snail. If the snail is left out, the little cosmos does not thrive so well, nor survive so long as it would otherwise.

Within the open pond or stream the same general relationships hold, but of course with many elaborations of detail. Plants of all degrees and sizes are present, and animals both large and small. Everywhere is the most intense activity, and the chain of food involves all of the organisms present. The largest fish feed upon smaller ones, these in turn upon the smallest. The smallest depend upon insect larvae, tiny crustaceans, worms, and other minute creatures that feed directly upon the microscopic plants. The larger plants, whether rooted or floating, are also exceedingly important, in the fresh and in the partly decayed condition as well.

To maintain this system of life, sunlight is absolutely essential, for the energy from the sun is utilized by the green plants, large and small, both to manufacture the food and to release the oxygen, which together maintain the animals. The clearer the water, obviously the more the sun's rays can penetrate beneath the surface on their beneficial mission. Yet steadily with the settlement of the country the water has become murkier; all of the forces that accelerate erosion have muddied the waters. The rainwater that used to be screened and filtered through layers of moss, leaves, and grass before it reached the brooks now scours into them its load of filth direct. As the surface humus has been washed away, a greater proportion of clay has gone from the subsoil into the streams, there to hang suspended after each rain like slime in a tumbler, screening out the sun for long periods with wet cloudiness. In what was probably a misguided moment, the German carp, long raised as a domestic food fish in Europe, was introduced into the waters of this country. It spread with great rapidity and flourished in the fast-muddying waters of the United States at a time when game fish were rapidly decreasing in number. Palatable when smoked or given other special treatment, carp is not highly regarded as a food fish in America. More a waterhog than a game fish, it has been given blame that should be shared for ruining our streams and lakes. Beyond question the carp is a scavenger, eating the eggs of other fish, rooting out aquatic vegetation and stirring up mud. Yet, like the Russian thistle in the western range, it is likely that he should be regarded as a symptom, not the real culprit. The increasing muddiness of the streams gave him his great chance, which, naturally, he took. To feed, in our present muddy waters, the normal population of game fish that the waters once supported would be as easy as to feed the human beings in this country entirely from crops grown indoors, with the sun shut away by walls and roofs.

Life in the waters must have its fertilizer, just as crops that grow upon the great farms on land. In the tranquility of nature undisturbed, this comes in the gentle drainage from the fertile soil of forest and meadow, so that the waters are rich enough in the various chemical elements that are needed for the food manufactured within them. There is a story of a small lake, in one of the northern states, that yielded fabulous returns to the fishermen who lived around it. So seemingly inexhaustible was it that the villagers found it profitable to abandon their other sources of income in favor of fishing. This situation finally attracted the attention of a man who knew a great deal about water and its ways, for no lake in the region had such a record. He found the answer. The village fathers for a century past had been laid to rest on a knoll overlooking the lake and draining into it. Transformed into lime and phosphorus, nitrogen and potash, these genial old gentlemen of an earlier generation were seeping down into the calm waters below and enriching their descendants without devise or inheritance tax!

In one of the southern states a city reservoir that had enjoyed notably good water began to change chemically, and physically as well, becoming much less desirable. The reservoir lay in a huge bowl covered with native prairie. Investigation showed that instead of being pastured or mowed, the prairie was burned over frequently, thus destroying its helpful filtering action on the rainfall and releasing an unbalanced chemical mixture, as well as undesirable sediment, into the water of the artificial lake below it. Manifestly what is good for the soil itself is good for the waters that drain off of the soil. *A watershed that is properly managed from the viewpoint of the land is the first step toward well-managed streams, lakes, and reservoirs.*

Through geological time streams have carried loads of mineral particles, dropping the heaviest first, finest (silts and clays) last. Many factors influence the rate of this process, but in general the presence of a cover of vegetation tends to restrain and regulate it. Wherever human numbers have increased beyond the capacity of hunting and gathering to support them, the new ways of life needed for survival have drawn heavily upon this living cover. That cover, in addition to protecting the earth's surface, converts it into soil by living in and on it. Unprotected, the soil becomes vulnerable to moving water and air.

In North America a few centuries have seen the changes that took millennia elsewhere. Seasonal floods have always been normal; the flood plain of a river is exactly what its name implies. But with clearing, plowing, and the building of roads, each road a new drainage-way, water began to move faster and pick up greater loads. Its scouring power is quadrupled as its speed doubles. Streams once clear became muddy and muddy ones opaque, blocking the power of sunshine to support aquatic life.

Meanwhile, the industrial revolution was gathering momentum, while speed of travel and transport grew with it. Immigration added its tens of millions to those resulting from high native birth rates. Thus the wastes of industry and human living joined the mud in streams already overburdened. What had not too long before been productive and delightful streams became free and loathsome open sewers. A single generation saw the intake crib for domestic water from one of the Great Lakes move out from near shore toward the center of the lake as the murky border of visible pollution spread away from the land.

The fault for our mistreatment of waters, as of forests, soil, and mineral deposits lay less in ruthless greed than in our belief that the vast richness of the North American continent had no limits. Where resources, including space, were scarcest, the story was different. In 1891 a young metallurgical engineer, member of a family notable since colonial times, visited the great German industrial center of Essen. This was at a time when that highly disciplined nation was becoming a world leader in many respects, almost a quarter-century before her tragic attempt to consolidate that leadership by arms. Surprised to find the Ruhr river clean as it entered Essen and clean as it flowed away, the young American was assured by his hosts that they could not afford to have it otherwise; that they had seen the condition of rivers in the northeastern United States and could not understand it on simple economic grounds.

Despite the environmental interest aroused by Theodore Roosevelt, and through its subsequent lapse and revival in the 1930's, only two sources of protest have had much initial effect upon our treatment of inland and coastal waters. Where evidence of waterborne disease such as typhoid could not be ignored, and where the wrath of sports fisherfolk became politically important, action followed. Claims for aesthetic decency, olfactory and visual, trailed far behind, as did those of the kind of long-term economic health that makes ecological sense.

The first measures were simple and direct, but had nothing to do with basic remedies. We are indebted for the chlorine cocktails coming from our faucets to the concern for public health. And we owe to the political clout of frustrated fisherfolk the elaborate system of hatcheries and restocking, providing a tight schedule of put and take. Reassuring, however, is the growing, but far from perfected, system of filtration, sewage treatment, and development of uncontaminated sources for domestic water.

Some notion of the situation comes from a five-fold increase in reported fish kills between 1970 and 1974, a fifteenfold increase in the damage caused by municipal pollution, and a tenfold increase in the kills reported for estuarial waters, essential spawning grounds for important seafood. Curiously enough, the kills reported as being caused by industrial pollution decreased during this period; much credit goes to an increasing sense of responsibility on the part of industry along with a concern for good public relations. And for the minority of industrialists immune to these influences, the fact that fish, like wildlife, are state property and their destruction is a tort against the state for which damages may be assessed, has had its effect.

Factory wastes very frequently contain the salts of heavy metals, particularly copper in solution. No schoolboy needs to be told that copper is poison. But sensitive as the human animal is to copper, the microscopic green plants that live in water and that are absolutely necessary to feed the fish, are infinitely more so. Less than one part in ten million will prevent their growth. This fact is often applied to keep them out of reservoirs where their presence is not desired. A sack full of copper sulphate crystals is towed around in the reservoir at infrequent intervals behind a rowboat, effectively discouraging this kind of vegetation. A river that receives the waste of any factory working with metal is likely to have in it a good deal more than necessary to sterilize it against the tiny plants on which the fish are so dependent.

Both factory waste and sewage may contain a good deal of oil. Now oil is an excellent thing to kill mosquito larvae in stagnant water, for it covers the surface and excludes the air that they need, in addition to coating them with a fatal film when they wriggle to the surface to breathe. The blanket of oil is no less effective in stifling fish by preventing the necessary interchange of gases with the air, particularly when other conditions have cut down the amount of vegetation available under the water to purify the air that is below.

How about sewage itself? Containing as it does a great deal of organic material that might be used by plants and animals for their nutrition, is there any real basis for objecting to it, aside from the aesthetic one? Certainly sewage is rich and concentrated. Civilization cannot go on indefinitely without utilizing it to a much greater extent than the western world has ever done. But when dumped into waters that do not have sufficient capacity, it undergoes rapid fermentation and decay. These processes require a great deal of the oxygen so essential to the animals in the water and release a surplus of the carbon dioxide that is so injurious to them. The effect is like that of great overcrowding. The sewage, so to speak, competes with the fish for the limited air supply. Additional damage probably comes from the waxy residue with which the waste finally coats the bottom of the stream bed, producing there bad biological conditions for plants and animals alike. Actually much can be done to make sewage safely usable. Further research and action on this problem deserves high priority.

Through the failure of great metropolitan areas along the coast to dispose of their sewage and waste in a sanitary manner, extensive areas of oyster beds have been rendered unsafe. The oyster is an excellent host to the bacteria producing typhoid fever, as many people have learned to their sorrow. Wherever there is the slightest danger of contamination, therefore, the use of shellfish from the questionable area is, and should be, prohibited. The area is thereby rendered economically, if not actually, unproductive.

Man's debt to ocean borders goes far back in prehistory, attested to by the shell heaps known as middens that abound throughout the world. The brackish to salty waters ranging from the estuaries of single rivers to great bays like the Chesapeake have been rich in such harvestable life as crabs, clams, oysters, and fish. The science known as marine biology goes back at least to the time of Aristotle. But only in modern times has it revealed the extent to which coastal shallows serve as nurseries for the great offshore commercial fisheries. Added to this, as pothunting has given way to more sportsmanlike harvesting of ducks and geese, the marshes fringing these waters remain a major source of feathered wealth.

Yet the natural generosity of these shallow waters continues to be ruthlessly abused by the pressure of human numbers and by those ways of modern life we call free, democratic, enterprising, and civilized. With New England marshes and shallows reduced to less than 20 percent of their original area, Massachusetts has been among the first to attempt to control this costly change. Elsewhere, and most notably the Chesapeake Bay area, the growth of cities, resorts, ocean shipping, and vacation homes has vastly increased the load of raw sewage. Countering this, in promise at least, is the louder rumbling of concerned protest that is now politically audible.

Although Americans eat a considerable amount of fish, and possess extensive commercial fisheries along the coast and on the Great Lakes, throughout the interior that form of food is regarded more as an accessory than as a staple. Rob an American of his coffee or tobacco and you have a rebel on your hands. Take the fish from his diet and he scarcely misses it, even though he be the worse nourished without it. As a consumer of fish, the American public is decidedly apathetic. As a catcher, either professional or amateur, it is very much in earnest and has obliged public agencies to give the matter of fish culture their serious attention. We have in consequence great hatcheries that annually return to the waters countless numbers of animals. As with salmon, such methods have enabled this unbelievable industry to continue on a machine-age basis. In mountain areas where the waters are relatively undisturbed and fishing very heavy in proportion to stream size, the sport of fly-fishing has been kept going. In certain privately owned and regulated waters, too, there is excellent fishing, by virtue of heavy stocking and restricted catch.

Not so with the great bulk of inland fresh water, into which carloads of young fry are released annually by state fish and game departments. The returns are meager and disappointing, more often than not negligible. The trouble is not due to persistent and unlawful fishing, even though there is still today a considerable amount of illegal seining, poisoning, dynamiting, and other refined kinds of sportsmanship, which our haphazard system of game wardens cannot cope with. The real trouble is with the water into which the carefully reared young fish are released. Murky, unbalanced, sterile for plants as we have seen it to be, rotten with decaying waste, it is in truth a watery desert. To release millions of fingerlings into such a mess and expect them to live is a priceless absurdity. It has exactly the same logic we would find in driving a herd of yearling Shorthorns into Death Valley or pasturing them on a concrete highway. If the fish and game commissions of our several states are to be more than agencies for the useless collection and disbursement of funds, they must be given authority and granted cooperation much greater than they now have. The hatching of fish, no matter how scientifically done, is but the beginning. Fish, like chorus girls, must eat. And, as we have seen, this involves a more favorable set of conditions than agriculture, industry, and civic development have brought about in our inland waters.

If we could rob the term "fisherman's luck" of its present melancholy implications by making our streams once more wholesome and beautiful, the world would be a better place than it is, both for the fisherman and those who have to live with him. (While the problems of fish and wildlife administration remain grave in 1979, the quality of that administration has steadily risen. Particularly important is the leadership of enlightened sportsmen in resource conservation and management.)

13
It Must Be the Weather

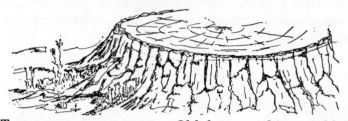

THE BOUNDARIES BETWEEN Oklahoma and its neighbor states were regarded rather casually until the discovery of great basins of oil. Thereafter a considerable range of talent was employed to settle the matter. At one place the age of trees growing in an abandoned river channel became important, since midstream marked state limits at the time of the original survey. An expert from the north central states, counting growth rings in these trees by taking cores to their centers, found enough of these rings to convince him that the channel had been abandoned long before the first surveys, thus giving the state that had employed him a generous slice of the disputed land.

His satisfaction was rudely ended when an opposing expert, native to the Southwest, demonstrated that there were often two growing seasons in one year, separated by a rest period during summer drought, as the familiar annual rings elsewhere were set apart by winter.

In lighter vein, but not without a point, is the allegedly true account of one of the scientists engaged on this problem of geographical research. Stopping at a cabin for a drink of water, he was queried by its presiding matron as to his business. When he explained somewhat jauntily that he was trying to find out whether she was living in Oklahoma or Arkansas, she exclaimed (he soberly insisted), "My God, mister, I hope you find we're in Arkansas. They say that Oklahoma climate is terrible."

There are places in the world where conditions vary so little that conversation about the weather would be considered inane. But the United States is not one of them; it has even been suggested that the familiar term "temperate" be changed to "intemperate" to describe the climates of this great country fairly.

This is not surprising; what we call weather and its more general pattern that we call climate are the result of the behavior of a fluid envelope surrounding a whirling sphere receiving energy from the sun. That energy falls upon a varied surface of land and water and with intensities that vary, not only with the seasons, but also with conditions in the atmosphere and at the surface. We are properly reminded that we are living on the bottom of an ocean of air!

Nor should we be surprised to learn that there have been, and still are, the weatherwise on land and sea whose intuitive judgment is not to be taken lightly. Successes of the Indian medicine man in invoking rain are attributed by skeptics to the fact that he alone is judge of when to perform his rites—failures not of record. Actually, although the attempts to discuss weather scientifically go back some three hundred years B.C., little progress was made until the seventeenth and eighteenth centuries, when means of measuring temperatures, pressures, and the behavior of gases and liquids were devised.

These measurements made possible the accumulation of records during the nineteenth century, mostly at the earth's surface. Then, following the two great world wars, observations were extended to the upper atmosphere, giving a three-dimensional picture of its activities. With this information, the ability to predict has been greatly increased. Yet it has far to go as current news attests: school children in Florida killed by tornado, the streets of New Orleans under deep flood, and Chicago paralyzed by heavy snowfall.

Sir Hubert Wilkins, known for his polar explorations from the air and under the sea, felt obliged to explain that his lifelong interest was not a mere whim. As a lad in Australia he had seen sheep on his family estate die from drought. Seeking the cause of such disasters, he learned that the Antarctic is deeply involved in Australian weather controls; hence, a resolve that guided his career.

The ceaseless search for general principles has extended beyond recent and current information about weather and climate. When Louis Agassiz demonstrated in 1840 that glaciers had not always been confined to mountains and regions now cold, but had at one time spread over continental lowlands, it became clear that climates and climate changes of the past were an essential part of the record. With everincreasing refinement the search for evidence continues, drawing upon the study of microscopic fossils in old lake beds and upon advances in the physical sciences. For example, we now know that the last great ice front (of four during the past million years) was some eighteen thousand years ago where Cincinnati now stands.

We also know that both the advances and retreats of these great continental ice masses were pulsating rather than steady, like the footsteps of a walker in slippery mud. The Great Lakes area was cleared of retreating ice about ten thousand years ago, not too much before the beginnings of agriculture in more congenial climates. Since then a warmer, drier interlude allowed such plant communities as grassland to shift northeastward along the great climatic diagonal that extends from the deserts of the Southwest to humid New England. More recently, a warming that melted back Norwegian glaciers after 1870 is credited with a shift northward of important Atlantic fisheries and a similar movement of some animal species on land. There are some indications that this change is beginning to reverse itself, but how much and for what length of time cannot even be guessed.

As knowledge of the remote and recent past, along with current observation, has increased, so has the search for causes and regularities that might make prophecy possible. The present, commonly thought of as postglacial, has shown enough similarity to what is known of interglacial periods to justify belief in the possibility of a return of glaciation in the remote future of human experience, if man survives to witness it.

Water for the formation of glacial ice was evaporated from the oceans, whose level fell about three hundred feet as a result. Increased intensity of radiation from the sun seems the most reasonable explanation of this fact, with enormous snowfall where snow still falls.

Accumulating faster than melting, changing to ice of immense thickness and weight, this frozen water from the sea ground its way to lower levels. Carrying in its path what it picked up, it provided the fertile mixture of limestone and shale that has enriched the north central states and the less productive loads of sand and granite particles that form much of the New England surface.

The cold of glacial times that forced the migration of plants and animals may well have been the result rather than the cause of huge formations of moving ice. Certainly it seems impossible to rule out the importance of changes in the amount of solar energy reaching the earth. And since the sun shows an average maximum of sunspots every eleven years, the meaning of this phenomenon has been pursued with the zest given in earlier times by alchemists to their search for the philosopher's stone and by astrologers to the riddles of human nature and destiny.

For comparison there are now available not only the imperfect measurements of less than two centuries, but another source reaching back well into the Christian era: the growth rings of trees. Simply put, these rings, especially those formed in critical locations, are wider from greater growth during years of ample moisture, narrower in years that are drier. Like sunspots, they show an interesting periodicity; wide rings tend to come in groups, as do narrow rings. Unfortunately, these groupings vary considerably and have to be studied, as do the sunspots, as averages.

The result has been some claim for the existence of eleven-, twenty-two-, and five-and-a-half-year cycles in the pattern of moist and dry groups of years. We have, however, interesting comments by the late, great geographer Warren Thornthwaite, who had mathematical skill along with his other talents. Not only did he demonstrate that during the extreme drought years of the 1930's there were locales of more than normal rainfall *within the afflicted area,* but that cycles of varying length could be derived from the raw data, depending on the assumptions used in calculation.

The great difficulty about reaching a useful working understanding of the weather and climate is in the matter of perspective. The weather of a day, a year, or a decade, is only part of the story. Just as the sharp edge of a razor resolves itself into a matter of notches and nicks, each with its own, smaller irregularities, when we look at its magnified image, so with the behavior of our atmosphere. A particular rainy day may be all-important to the man wishing to set out tomato plants, or to another interested in the outcome of a particular horse race. It would be of little importance to a cattleman, whose pastures depend upon the total annual rainfall and its seasonal distribution. And the rainfall for a particular year and place would not loom very large to an operator whose activities were so extensive and diversified that loss at one place or time is counterbalanced by gain at another. And for the last named, only a major, protracted shift of climate, entirely changing the aspect of a region and its cultural life, would be of decisive significance.

Climate depends upon the interaction of a great number of factors. In consequence, its manifestations cannot be described by the method of individual portraiture, any more than the American people can be so described. It must be dealt with by means of statistics, and when this is done, we are rewarded by an understanding that, if not complete, is at least very useful. Coming back to our razor blade, no amount of study would enable us to draw in advance a precise picture of the microscopic nicks that would show in a particular blade. But it would by no means be impossible to estimate in advance the size and number and average depth of them for a blade of known material and shape, sharpened by a certain process. This is the sort of useful knowledge about weather and climate that we have, even though we have not mastered the art of using it.

Climate has been called "the average weather," not a strictly accurate term, but helpful perhaps. What does this involve, using the comparatively simple matter of average annual rainfall? Over a long period of years this may average for a given place forty inches. But the actual years with precisely forty inches, no more, no less, will be few if any. The greater proportion of years will be close to that figure, of course, with extremely wet or dry years fewer in number. There seems to be a definite tendency for the drier years to be grouped together (likewise the wetter), but not in such an invariable way that one can say, "next year will be dry" or "three years from now we shall have very wet conditions." Even though there is evident a rough tendency for the cycle of wet and dry to complete itself about every eleven years, we can only prophesy with the wisdom of the sage, not with the occult power of the soothsayer. We know, with the inspiration of the old Pennsylvania Dutch farmer, that "it never rains much in a wery dry time." And that no year is going to be precisely normal we may be sure. There is this much control: statistics tell us accurately the extent of the hazard in either direction that we must be prepared to run.

Now that is as much as life insurance companies have in the way of information upon which to base their very successful operations. They cannot tell a given client when and how he is going to die, but they know a great deal about the likelihood involved in this event. If many who are insured live longer than they are entitled to, enough fall short of the mark to compensate for that fact. Such is the power of prophecy conferred by statistical analysis.

How can this be of any practical use to the man who works the soil? In the first place, in making his capital investment, he has in the past too often done it on the basis of bumper years. Two or even three such years may follow in succession because of the tendency we have mentioned for years of high rainfall to occur in groups. Needless as it may seem to mention that such conditions are never permanent, the sad fact remains that countless investors have acted exactly as though they thought otherwise. Not only will the majority of years be less favorable, but years, and probably groups of them, must come when conditions will be decidedly unfavorable, even disastrous.

As long as bankers permit the farmers whom they serve to disregard these simple facts in purchasing land, this branch of industry will be in the position of an insurance company that extends protection to clients of any age, figuring its premiums on the life expectancy of a person just turned twenty-one. Perhaps it is naïve to suggest that anyone who really wants to help agriculture on to solvency can do no better than shout such facts from the housetops. But it certainly would be more naïve to expect those who stand to profit by the financial vicissitudes of agriculture to give the situation much publicity. There is considerably more to scientific agriculture than the production of two blades in place of one. Almost any fool can do that when nature is with him. The goal is a system that can stand the shock of years when half a blade is all that can be had. And as a sound beginning, no more hard-won capital should be involved than the inevitable turn of the cycle will justify.

Nor is this all. Civilized man ought to be less the sport and victim of the elements than he is. Good husbandry enhances the benefit of favorable years. It can, almost equally, stand as a buffer against the certain impact of unfavorable ones, and should have that in mind as a conscious end. But it must be remembered that the two ends are not the same, and the desire for the first should be tempered by the thought of the second.

Centuries ago a royal Pharoah discussed this problem with an alert young Hebrew named Joseph. Joseph advised him to lay by the surplus of the fat years to tide over the lean. This advice is still as good as it ever was. And it must be conceded that the advances of modern technology have done their share toward this end. Vast storage facilities are now available, and methods of preparing materials to insure their long keeping. For example, the discovery that wheat flour will keep fresh much longer if the germ is removed in milling has certainly not helped the nutritive properties of the flour, nor its flavor when fresh, but it has made possible long shipment and a great spread of availability of this staple commodity. Canning and preserving suggest themselves at once as means towards the same end. In any criticism of the present system of private initiative, it must be borne in mind that these advances have been largely developed and made workable under that system, with tangible benefit to the whole social order. Unfortunately, private initiative has not yet suggested the means by which, in times of dearth, those who are without purchasing power can share the stored resources of rich seasons—unless we regard installment buying as a not altogether happy gesture in that direction.

But the question of lean and fat years goes deeper than the problem of reserves and equalizing consumption. For answer here we must look not to the good sense of a Joseph, but to the spiritual admonition: "Consider the lilies of the field, how they grow." And in truth, they have the answer we are seeking. In the prairie states there are many kinds of introduced ornamental and fruit trees, in addition to those that are native. The change from the frequently open winters to spring is a vacillating, disorderly process. Benign and lovely weather in February or March may be followed by lethal frost; growing days are interspersed between times of howling blizzard. The usual result is that the introduced plants are enticed out of their winter safety, putting forth leaves and flowers only to have them trapped and killed when savage cold replaces balmy warmth. So usual is this that in some parts of the grassland states—where peach trees grow very well indeed and produce excellent fruit if the blossoms are not killed—a good peach crop about once in every four or five years, or about three times in the life of the average peach tree, is the best that can be hoped for. The peach, of course, is not a native fruit.

Meanwhile, it is notable that the native plants are seldom roused by any false alarm of spring. Cottonwood, redbud, oak, and wild plum usually remain quiescent until the period of nature's caprice is past and then emerge in safety. Why and how this happens no one knows, but of the benefits of it there can be no doubt. Somehow or other in the course of an immeasurable past, Nature has eliminated the unsuitable in such a way that the flora of each region is remarkably adjusted to local conditions. She may proceed with great leisure about this task. We know that oranges grew for 140 years in Georgia before they encountered a killing frost. Thereafter they were quite as dead as though they had been eliminated in the first 5 years. Much of man's trouble in the lean climatic years is caused by the fact that he is forever attempting to grow species or varieties of plants in a climate to which they are not certainly adjusted. Particularly is this disastrous when the commitment is heavy—when all the eggs are in one basket. The remedy does not need to be labored in the telling.

Suitability of crop to the hazards of climate, then, is our first lesson from the lilies of the field. The second is no less important. Nature abhors monotony as she does idleness, unbalance, or vacuum. She clothes the native field and forest with variety, as we have seen, for example, in the virgin grasslands. We have observed there how the mixture of native plants possesses a resourcefulness and flexibility equal to any of the ordinary vagaries of rainfall, temperature, and evaporation. Conditions one year may favor a certain part of this mixed population; the next year a different constituency will form the favored group. However the pendulum may swing, the round of life goes on.

It would be unfair to say that tillers of the soil have not utilized this principle. In the western half of the prairie states, where the threat of dryness is never far off, it has become a part of standard practice to keep at least a small acreage in the drought-resisting sorghum, or "cane," planted so thickly that the stalks are slim and spindling. Although far from a perfect ration, this can be used in place of hay if all other crops fail, and has pulled more than one farmer through the winter following a bad year. But in comparison to what should be done in the way of diversity of crops, it represents a mere gesture. The hazard of the climatic cycle cannot be minimized until a variety of crops becomes the invariable rule and animals are restored as an essential part of the economy, not only for their marketable products, but their wastes so necessary to the soil.

The third lesson from the lilies of the field lies in their effect upon the soil. Both they and the animals that in undisturbed conditions eat them return each year's surplus of new organic material to the ground, depositing on the surface minerals brought up from below, and creating humus that keeps the surface moist and mellow. Further, this surplus of litter keeps down to the utmost limit the amount of bare space and holds the ground against washing and invasion by weeds. Thus the soil itself is not only kept in the best possible physical condition to withstand unfavorable seasons, but is continually being improved. By contrast the usual result of cultivation, as we have seen, has been to produce physical and chemical deterioration, leaving the soil and the plants upon it almost defenseless in the lean years. To mine the soil persistently and trust to heavy doses of chemical fertilizer when it becomes exhausted is to disregard the plain and explicit warning of nature.

The final lesson that the seeing eye may glean from the plants of the field is that they themselves represent one of the steadiest and most certain sources of wealth, in lean years as in fat. They thrive under moderate and balanced use and are ever ready to yield their quota of material to the mower or the shepherd, provided he does not abuse this generosity. The farm or ranch with a goodly proportion of its surface still in virgin sod need not fail, whichever way the cycle of climate may swing. By disregard of this insurance on millions of acres, citizens and nation alike have brought upon themselves the dire penalties of poverty, hunger, and discouragement.

Since crops and practices must be intelligently adjusted to climate, how can we map the climates of our country to guide us in our practice? It is easy to map the separate features of climate, such as sunshine, temperature, and rainfall, but alone these tell us little. We must have a way of expressing their combined effects so that we may know how the growth of plants is affected by them. Furthermore, it is very difficult to know how refined to make our distinctions. Just as there are counties within states, so there are climates within climates. We could easily go too far for practical use. Many discouraging attempts have been made to map climate in such a way that it would be most significant in terms of native vegetation and, correspondingly, for cultivated plants. In recent years ways have been found to take account of rainfall, evaporation, temperature, and seasonal pattern so that extremely valuable maps can be made. When these maps and their meaning are well understood, it will be possible to know in advance whether great enterprises such as the extensive introduction of new crops and methods represent a sound development or a rash gamble wagered against the loaded dice of climate.

Perennially the question arises: Is the climate changing? If it is, all effort becomes a gamble, of course, unless we can predict the course of change. It is true that there are abundant evidences that climate has not been uniform throughout the whole of geological time. Plants have grown luxuriantly in what is now frozen waste near the poles, enough so to form considerable deposits of coal. At other times vast icecaps have spread far beyond the present limits of the frozen polar regions, and the snowcaps of mountains far below their present lowest levels. But is the climate changing now? To begin with, we must eliminate the personal impressions that people form during a single lifetime. There is plenty of trustworthy record to show that in such a brief space of time only the fluctuations we have already discussed can have occurred. One of the most persistent opinions, even in scientific circles, is that climate is gradually becoming drier. Before this is accepted, we must grant that the effect of human interference with natural vegetation and soil nearly always takes the form of making rainfall less efficient and of introducing the weeds characteristic of drier climates than those in point. For example, burning and overgrazing encourage the cactus to move in in place of grass, and grass to come in in place of trees.

Mankind seems to have a stubborn genius for learning the most important lessons the hard way. The disasters of the 1930's, caused as they were by failure to respect the dangers of misuse in a region of known climatic hazard, were repeated later in Russia. Now, during the 1970's, shelterbelt plantations in the semiarid former dustbowl are being cut to provide more plowland, even while a consortium of nations, alarmed at the spread of deserts in various parts of the world, has met in conference to discuss this problem. One clear outcome of this meeting is the fact that desert expansion should rightly be called *desertification*—the enlargement of deserts by human pressures.

We have seen that man has been of two minds concerning the forest, his source of needed goods, but also his rival for space. In a different way he has shown the same ambivalence toward climate, ignoring its inevitable patterns, yet blaming it unjustly for his own follies. Early in the 1930's, an Oklahoman sufficiently prosperous to own an airplane returned from a flight to the High Plains, jubilantly announcing how many thousands of acres of shortgrass turf had been plowed up for wheat. Warned that the drought, already under way, was likely to continue, he scorned it as a bit of nonsense.

In northern New Mexico a wide valley stretches west from the Sangre de Cristo mountains to the deep gorge of the Río Grande. Below the pine and juniper wood-

land, the entire basin is covered with sagebrush except
for roads, buildings, and a few skillfully handled fields.
Under natural conditions this aromatic but inedible
shrub forms a narrow belt separating woodland from
grassland that occupies the valley floor. Such, actually, was the condition in this beautiful valley until
the end of the nineteenth century; since prehistoric
time, it had been a trade center for its foodstuffs. Then
around 1900 thousands of sheep were brought in, soon
destroying the grass and permitting the sagebrush to
move downward and replace it.

Yet old-timers are reluctant to admit that overgrazing had anything to do with the shift from grass to
sagebrush. Instead they blame the weather—a series
of dry years that does show clearly in the record. Ignored is the fact that grass had persisted for millennia
under the grazing of wild animals that were free to
move about, even after the arrival of the Spanish with
their horses, burros, and cattle. Not until the native
pasture was overloaded with thin-lipped sheep at a
time when it should have been used with caution was
the grass cover destroyed. As in most resource management problems, the answer is not a simple yes or
no, all or none, white or black.

And herein lies the really vital contribution to
planned economy that can be made by the science of
climate and weather. Interesting as it would be to en-
vision the climatic shifts of the next few thousand
years, and important as it is to have daily forecasts,
the real need is to impress upon the popular and official mind the fact that variation is the rule and not the
exception. Yet it is equally important to understand

that this variation between wet and dry, hot and cold, forms a reasonably orderly pattern to which human enterprise can and must be adjusted.

Once the character and extent of the inevitable fluctuations are appreciated, climatic hazard ceases to be the irrational thing that it is at present. Not until then will it be possible for land utilization to be elevated from blind and futile makeshift to a reasoned and orderly policy.

14

Rivals

OUR PLANET EARTH is, by the best available evidence, between four and five billion years old. For at least half that time it has supported some form of life. But never, until the rise of modern man, has a single species, our own Homo sapiens, been dominant over most of its surface, with the power of a great natural force. Yet the rule of mankind is not unchallenged. Challenge comes from other forms of life as well as from within the human race itself.

In the days before the Department of Agriculture became heavily involved in matters of economics, sociology, and high statesmanship, the Bureau of Entomology was long headed by the distinguished L. O. Howard. Among his many valuable services to the nation was his reminder that the last living thing on earth might well be an insect on a dead weed. This of course was no denial of the immense value of insects to man, which no one understood better than Howard. Aside from such products as honey, wax, silk, and dye and their invaluable work in the pollination of useful plants, they serve us in many ways, some understood, some partly, and some probably not at all. What Howard meant to dramatize was their immense biological success.

The number of known insect species is approaching a million; how many others remain to be identified is unknown, but enormous. Their variety, capacity to multiply, and programmed patterns of behavior all contribute to make them man's most effective rivals for the food supply of the world. The tiniest of insects contains within itself the basic means of survival as truly as does the elephant. Miniaturization is no novelty developed by modern electronics, amazing as that may be.

At every step in the production of food and fiber, as well as in its storage, processing, and distribution, these invertebrate rivals take their toll in an aggregate that is staggering. Nor are insects our only rivals for the organic material that we need. Rodents and other destructive animals, as well as microscopic plants that cause disease and decay, are unbelievably greedy. Our efforts to relieve hunger in undernourished parts of the world are handicapped by failure to protect our shipments of grain from rats and mice. Some estimates of loss run as high as half. A hundred rabbits are said to eat as much as one cow, if not more. Weeds, while manufacturing their own food, do so at the expense of space, sunlight, and water, all of which are needed by useful plants. The annual water requirement of a sunflower plant is at least equal to that of the corn plant growing in a row—all too often beside it. And since much of our foodstuff must depend on rain that falls before the growing season begins, we may see how little there is to waste. In the grain belt as the roots of either weeds or crops grow downward, they progressively dry out the soil in their course. Only in unusual seasons do the summer rains really renew what has been stored and then removed. All along the line it is war to the knife.

The chronicle of this unceasing battle is one of absorbing interest, no matter whether one can understand the technical details, or whether he is simply interested in the adventurous angle of it. A Frenchman, spraying his grapes with blue vitriol to sicken thieving boys, finds that the fruit is unharmed by disease that ruins his neighbors' crops, and thus invents the first fungicide. French peasants use Ku Klux Klan methods and by direct action destroy the government hedges of barberry, which, they say, are blasting their wheat with rust. The government, thus moved to secure scientific counsel, finds that the peasants are right and the long war against the rust begins.

California citrus groves are threatened by scale insects that secrete a poison-proof coat of wax, beneath which they suck the juice of the plant with impunity. The lady beetle is found to be exceedingly fond of scale insects and is marshaled into an army that does the work as neatly and completely as any mercenaries ever served a royal crown. In today's dubious paradise of labor-saving devices, it should be a rule never to try to do mechanically what you can get another form of life to do for you. The effort to find parasites fatal to sugar cane pests was under way as early as 1910. Reports from Hawaii now indicate that the yield from unsprayed cane fields is more dependable than that of plantations given a lethal baptism; poisons fatal to our insect enemies can do away with our friends no less.

Considering the odds against us, our survival thus far is remarkable, while its permanence can by no means be assumed. We may score a success by sterilizing enough male insects to prevent the production of fertile eggs by females whose mating instincts they satisfy. But we may find that although we bait traps with a favorite attractant of another species, it has little effect. Simulated alarm calls may scatter flocks of unwanted birds, but with them as with the humans of Aesop's fable, the warning soon loses its impact.

Our most notable defeats have come from the rapid spread of rivals—parasites, predators, and nuisances—introduced from other environments. Recent examples include the chestnut blight spread by birds and Dutch elm disease, by bark beetles. Of longer standing are the house sparrow and starling, rabbit and cactus in Australia, codling moth (source of wormy apples), and corn borer, unwittingly exchanged from Europe for our gift of potato blight and grape mildew.

By no means do all of our troubles come from invading rivals. We can foster our own. Our western grasslands are home to numerous kinds of rodents such as jack rabbits, prairie dogs, and gophers. These live on the same vegetation that supports grazing animals; they also destroy the native sod by their burrowing. However, they furnish food to such animals as the coyote, a circumstance that keeps the rodent population within bounds until we enter the picture with our flocks and herds. Whether it is literally true that a hundred rabbits will eat as much as one cow, where from ten to thirty acres is required to support one head of cattle, any added drain becomes important. And the heaps of earth around burrows encourage the seeding in of unpalatable weeds to replace the lost grasses.

Complications arise when the prairie wolf or coyote varies his normal diet of rodents, insects, and wild fruits with an occasional lamb or calf. Although much of the resulting damage comes from individuals that have learned to share our tastes, and a great deal from uncontrolled dogs, it is easier to stir up the martial spirit than to seek calmer measures. Where such devices as baited cyanide guns are used, the coyote population can be reduced to a point where rodents have a field day. Weeds multiply, and frequent overgrazing by cattle or sheep hastens damage to the range. There appears to be a kind of toboggan effect as this happens, with injurious organisms multiplying, temporarily at least.

In dealing with trouble of any kind, our first impulse is to treat symptoms; often this is the best we can do in emergency. But the history of medicine shows that the ancients knew a great deal about symptoms and had, by trial and error, learned much about the treatment of disease. Nonetheless, medicine did not really progress until modern science revealed what we know today about the healthy, normal human body. In dealing with our rivals for food and other benefits of the environment, the same principle applies. In folk language, the long way around is often the shortest way home.

The long and patient search for parasites of sugar cane pests has been mentioned. At one point, according to a gray eminence among biologists, progress was held up for a year or so to allow a technical study of wing venation in the wasp family, this to insure that an effective ally would be imported and not a useless or even mischievous relative. Again, the notably efficient organization that serves plant and animal industry in the state of Kansas has taken the trouble to study coyote predation with care, reserving control measures. The finding is that the damage usually comes, like that from the big man-eating cats, from a single individual. The state now has an expert troubleshooter ready on notice to find the culprit and dispose of it, instead of spreading poison wholesale. The most recent reports confirm the success of this procedure.

We have heard much of the plague of grasshoppers, crossing the country like a wave of devastation and consuming every green thing in their path. Yet a fence of three barbed strands of wire has been known to stop them. In the Wichita National Forest is such a fence. On one side the herbage is heavily populated with various types of destructive grasshoppers. On the other side the species present are somewhat different, and the numbers very much less. Actually, of course, the fence served to prevent overgrazing. But the truly surprising thing is that the hungry pests did not occur to serious degree on the side with the large amount of potential food. Like scavengers and troublemakers who have no place in an ordered existence, they found their opportunity only after the natural balance had been practically destroyed. Thus, when man begins the downward course of destruction, does nature operate to accelerate the dizzy process. As the damage increases, so does the possibility of further damage.

It is well known that maize does not thrive commercially much north of the Canadian boundary, nor wheat too close to the Mexican line. The former is essentially a warm climate plant, the latter one of cooler regions. It has been found that both are much more susceptible to disease when grown near the limits of their temperature preference than when grown safely within that range. And, in general, any plant or animal far from its proper home must have unusually favorable conditions to reach a healthy maturity.

Parasites and predators, while a constant menace, are seldom completely successful in their destructive efforts under what we shall call normal conditions. Doubtless there have been instances of such complete success, but only at the price of extermination of the victimized plant or animal—an eventuality that can have had only one outcome for the destroyer. The earth is peopled by organisms that have been through such fiery tests for an unimaginable length of time. Hosts and parasites have survived only because they have developed the ability to live together.

When a form is transplanted into a new locality to which it is adapted, it is likely to spread with astonishing rapidity because of the absence of competition and accustomed enemies. There are many familiar illustrations of this, ranging from the English sparrow and starling in North America to the rabbit and cactus in Australia. Probably the red man exhibited the same sort of rapid population increase when he entered our continent, and certainly the white man has done so. European crops such as wheat, brought to this country, have behaved the same way under the protection and intentional encouragement of man, and so have the various American plants that were taken back to Europe, particularly the potato and maize. We know, of course, nothing as to how these invasions operated before the time of man's appearance, and particularly before he had developed to the point at which he became an important, conscious factor.

So far as we can tell, however, the invaders, whether plant or animal, are eventually compelled to slow down. If nothing else stops them, competition with their own kind will do it. The period of unrestricted increase ends and the newcomer subsides to his proper place in the picture. Just what this place will be depends upon many circumstances. In the case of the English sparrow the change in domestic architecture away from Victorian gingerbread to the stark simplicity of modernism has certainly affected the abundance of nesting sites. Better care of garbage and other waste, and the supplanting of the horse by the automobile have restricted the opportunities for feeding in the cities and towns. This bird, once described as an incorrigible, dirty gamin of the streets, has now become rural, and has had to go back to the farm to live on shocked and shattered grain that the shiftlessness of some farmers provides. The European starling with its noise and filth has not hesitated to desecrate even the capitol of the nation, arousing the ire of others who wish to be heard in that forum and defying the onslaught of those responsible for the maintenance of order and calm. Gas, water, and alarms all seemed to be ineffectual to do anything but provide copy for the newspapers. The bird likes it here and apparently intends to remain.

Man himself is by no means immune to the rule that

operates upon the successful invader. The American Indian, although he had developed remarkable centers of culture in Mexico and in Central and South America and had spread over the rest of the New World, did not by any means have a dense population. It is unnecessary to try to enumerate all of the influences that operated to limit his numbers, and indeed many of them we cannot know. But we are certain that he had serious technical limitations in the matter of tools and available power, and enough in known of the history of the Mexican civilizations to know that competition with his own kind was at times a potent, even a decisive factor. In fact, so delicately balanced was the adjustment within this comparatively sparse population that the pressure caused by the earliest white settlements on the Atlantic coast seems to have been transmitted to the interior, causing serious displacements there long before the first explorers had reached that part of the continent.

White man with his superior technical resources has been more successful, if numbers are a measure of success. But epidemics and wars have killed their millions, while famine has stalked the land, not always overtly, but often in the invisible cloak of malnutrition and prolonged underfeeding. The Civil War is not to be measured alone in terms of lives and property destroyed, but also in terms of the spiritual discouragement that for nearly two generations thereafter clouded the once-proud South. Energy and hopefulness, no less than the capacity to reproduce, are essential to mankind. And these were destroyed in the southern states to a degree that few people in the North have realized. Likewise, during the century fol-

lowing the Spanish conquest of Mexico in 1519, the
massive destruction of the native culture was followed
by apathy, frustration, and, if reports can be trusted, a
significant decrease in the native population.

Serious though the rivalry between different forms
of life may be, the greatest potential rivalry must be
within a species, man being no exception. For members
of a species have the same requirements for survival.
Whether desert plants, wolves, or songbirds, they usually meet these requirements by spacing; bare ground
between individuals is the mark of desert, sound and
scent are among the various signals used by animals to
mark limits.

The themes of history are largely concerned with
human rivalries. Ranging from friendly emulation to
murderous hostility, they command a high share of our
daily attention. Their resolution and control is one of
the heaviest burdens on government, exceeded only by
the cost of anticipating serious rivalries between governments, i.e., armament. Whether such a resource as
space is being fought over or open to unrestricted sharing (a form of rivalry), the net effect is generally disastrous to the resource and so, in the long run, to those
dependent on it. To this dilemma, which Garrett Hardin calls "the tragedy of the Commons," no satisfactory
political or economic solution is clearly visible.

As civilization becomes more complex, so do the
pressures on land, using that word as economists do, to
signify environment in general. Most obvious are the
rivalries between short- and long-term advantage,
aesthetic and economic claims, such as those involved
in the question of wilderness reserves. But even this is
complicated; ample natural reserves may safeguard

the future economy as well as the beauty of environment, while present economic advantage is too often a threat to that of generations yet unborn.

Although vast damage to the resources of the United States was done by a population less than half as large as that of today, human numbers remain a major source of pressure on environment. Fortunately, control of population increase is gaining ground where living levels are high and their cost proportionate. And though it lags in most places where most needed, it has sturdy support in China. *A Chinese Manual for Barefoot Doctors* outlines methods of birth control with a directness that might raise hackles in western communities that accept pornography without a protest.

Man, of course, has become the sponsor of a biological experiment without known parallel in the history of the earth and its inhabitants. He is the first example of a single species to become predominant over the rest. Heretofore, there have been great groups of species—invertebrates, fishes, reptiles, and finally mammals—each in its turn replacing those that had gone before and becoming the conspicuous, characteristic type upon the earth. As these newer and more successful ones appeared, the less efficient and older either became extinct or were relegated to a place where they did not suffer from more efficient competition. The reptiles of today are a modest and innocuous lot compared with those gigantic forms that once roamed, and claimed, the earth. But never, as far as we know, has a single kind so completely swept all others aside and taken possession as has mankind.

Furthermore, man has not only become predominant, but as the price of his power, indeed of his survival, has assumed conscious, deliberate control. He has established a new order, with his own good as the criterion of it. He is attempting to rule the earth as a god might do, not only seeking what he needs, but manipulating all that is around him, supplying the conditions of life for the lower organisms that he uses and combating those that are hostile with resources they do not have. He no longer accepts, as living creatures before him have done, the pattern in which he finds himself, but has destroyed that pattern and from the wreck is attempting to create a new one. That, of course, is cataclysmic revolution.

We have seen something of what this entails in the way of hazards and failures. Soil has been exhausted or changed, forests and grasslands destroyed, topography injured, and conditions produced that facilitate the activities of rivals competing with man for survival. Man is not ignorant of these consequences, but as an individual he acts as though he were. So vast is his empire and so numerous his brethren that no one feels, in any convincing sense, that his personal activities can constitute a serious menace in any but an immediate, direct, and personal way. So long as consequences of such measure are avoided, the individual takes little heed. The gift of intelligence with which mankind is endowed represents the collective effect of individual intelligence, acting in the main on problems of expediency. What develops in the way of policy seems the blind resultant of these innumerable integers.

The past century has witnessed some rather heroic efforts to break this fatal chain. In particular, wealth and science have been employed to combat the enemies and limitations under which man labors in pursuit of his daily bread. Engineering skill has devoted itself, often cumbrously and ineffectively, but in the main successfully, to relieving him of toil. He has delved deep into the mysteries of other organisms, with the object of utilizing or destroying them. There has been no stint of money for science, where the practical end could be demonstrated to legislator or philanthropist. Apart from the empirical wisdom that peasant tillers of the soil possess, and which has in Mennonite and Czech colonies more than once proved itself, it is safe to say that we have today not all the knowledge we need, but a great deal more than we use. That is to say, if what knowledge we possess were consistently applied in the tilling of the soil and the management of range and forest, the course of waste and destruction, exploitation and loss, would be arrested. Could that be done, the possibility of an ultimate equilibrium favorable to man would be immeasurably advanced.

It is needless to ask why this knowledge is not consistently applied. It is not, of course, uniformly diffused. Even if it were, it would not be as effective as it should. Character and self-discipline, the training that comes through apprenticeship, are all too often lacking in the individuals who, having the knowledge, must be entrusted with its application. This is the answer to those who cry for research and education. Both are good; both are essential. But they are not good enough. In all his study of the pests and predators that handicap him, man has not taken sufficient account of the chief handicap, which is himself. It is the human element in practice that is deficient. One may acquaint a young man with all there is to be known about sound practice, but if in early youth he has not acquired habits of responsibility, persistence, and industry, it is a waste of time. *It is not merely soil, nor plant, nor animal, nor weather that we need to know better, but chiefly man himself.*

15

Cold Figures

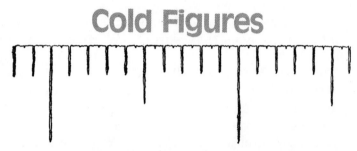

AN HONEST ACCOUNTING of public assets and liabilities has never been popular, nor is it now. Throughout history, as long as anything was to be had for the taking, within or beyond community borders, it wound up in the hands of those able to acquire it. A classic example is our own western world, incalculably rich, occupied at the time of European invasion by cultures unable, for technical reasons, to defend themselves effectively despite their often superior tactical skill. Presidential messages of the nineteenth century reflect continuing concern with the "Indian problem" along with evidence of a belief that our resources were unlimited.

Not until the politically unintended administration of Theodore Roosevelt in 1900 was there any major government effort to slow down reckless exploitation, and then chiefly that of the highly visible abuse of our forests. This president was our first to have a genuine knowledge of the West along with a considerable interest in natural history.

Earlier, and at enormous cost, we ignored an appraisal of the semiarid West based not on geometry, but on the natural landscape as its potential was indicated by topography, vegetation, and the water economy. Major J. W. Powell, intrepid one-armed explorer of the Grand Canyon, urged in vain that this vast empire be developed in functional units instead of the familiar rectangles that worked well enough in the humid East.

Theodore Roosevelt succeeded in reserving some eighty million acres of national forest before Congress tied his hands. Later, the impulse he had given to environmental concern was swamped by war and the period of economic euphoria of the 1920's. Not until the chastened mood following the debacle of 1929 was this concern revived, this time by President Franklin D. Roosevelt. Whatever his faults, he had the virtue described by one of his friends as a "willingness to listen." An important consequence was his formation of a Natural Resources Planning Board, set up to undertake a survey and analysis of all natural resources, population, past changes, present conditions, and future possibilities.

Actually, his predecessor, the professional engineer and humanitarian Herbert Hoover, had undertaken something of this sort in his concern for social trends. The training of engineers has often been criticized as too narrow, but it should never be forgotten that it involves a discipline in responsibility frequently lacking in other realms of higher education. A fledgling engineer is not allowed to forget that when he signs a blueprint or an estimate, it is neck or nothing. And while most engineers spend their lives working for others who decide policy, occasional ones like Arthur

Morgan and General U. S. Grant III are able to achieve broad perspective.

Whatever vision President Hoover may have had was obscured in the "Cloud Nine" atmosphere preceding the events of 1929. As survivors of that era can testify, common sense had been in eclipse. Anyone rash enough to question the plowing up of arid land or to suggest that the supply of petroleum had its limits was lucky to get off with nothing worse than being ignored.

Once established, the Natural Resources Planning Board failed to develop rapport with a Congress not too enthusiastic about the Board's mission. But before its demise, it produced an inventory and analysis of national resources. Also, it stimulated the formation of planning agencies at state level, while the brief activities of the board must be reckoned, along with other influences, as helping to increase the status of planning in academic circles.

The raw material for the Board's report was the amount and quality of the nation's area in relation to the number of human beings occupying it. For convenient recall, the area of the 48 contiguous states is somewhat more than three million square miles, slightly under two billion acres, 98 percent of it land, the rest water. Alaska, Hawaii, and miscellaneous territory increases these figures by roughly one-sixth, not only adding land-use problems, but also complicating, since 1950, the statistics.

Before this complication, the area of the contiguous states was reported as 1,903 million acres, a little more than half in farms (one-fifth of which was in crops). Forest occupied somewhat more than one-fifth of the whole, grassland not quite so large a share. Other uses such as roads, cities, etc. occupied less than 4 percent, while about the same amount was then considered useless. These figures took no account of whether land was being put to its best possible use, or managed in the best possible way. A great deal (for example, overgrazed grassland and cutover forests) failed to meet either of these requirements.

As of 1970, with 2,266 million acres to reckon with, less than half was in farms. Forest reportedly covered more than a fourth, with a sharp decrease in the percent of grazing lands and a tripling of land in "other" uses. Alaska, supplying the bulk of area not included in the 1940 figures, is now the scene of vigorous conflict as to future land use. Perhaps the most striking change in recent decades has been the reduction in the number of farms and an increase in their average size, made possible by the use of machines run by fossil energy (often in amounts approaching the energy yielded by the crops produced). At the same time, there has been a steady conversion of land to uses that remove it from productive agriculture—the annual amounts involved being of the order of a million acres or more. Much of this change can, so far as agricultural use is concerned, be compared to a lava flow, for it involves covering soil—often some of the best—with concrete or other structural material, a change that is irreversible in any practical sense.

The first census of the United States, taken in 1790, records almost four million people living on a little less than 900,000 square miles, or 142 acres per capita. By 1970, the individual share had shrunk to less than a tenth of that amount of space, a bit more than 11 acres, of which only about 4 are in farms producing food, for both domestic use and export, as well as cotton and tobacco.

We do fairly well in measuring areas on the earth's surface; we could do much better than we do now in surveying the quality and possibility of that surface. When it comes, however, to judging the future patterns of human numbers, we share all the pitfalls of the prophet. By 1820, the birth rate in the United States began a slow decline, as it did in most of the industrial world. The exceptions were in isolated rural pockets and in nonindustrial nations. Meanwhile, however, the U.S. population kept growing through immigration. Both urban life and improved levels of living tend to produce lower birth rates to such an extent that large cities must maintain growth or even stability by taking in those born elsewhere.

The effect of better living conditions is so clear that, as a sensible British biologist pointed out to agitators who wished to persecute a certain racial group, depriving it of privileges, the surest way to lessen the supposed danger was to give the undesired (not necessarily undesirable) group every opportunity to succeed and prosper. Under such conditions the birth rate inevitably must fall and the group diminish in numerical importance as time goes on. Yet there is the curiously contradictory fact that economic depression can act as a damper on fertility; people who have known a reasonable degree of economic security are likely to become acutely aware of the cost of added children when hard times come.

Unquestionably there was a notable decrease in births in the western world around 1921. Seldom noted is the fact that rural hardship struck at that time, eight years before the general industrial-urban panic of 1929, continuing until it reached disaster proportions in the 1930's. Doubtless the extent of hardship was obscured by the fanfare over mechanized farming ("factories in the fields") that contributed so much to the ultimate disaster.

Also involved was the power of the pen. Humanists and biologists were busy fighting the taboos that distorted or prevented a sane general understanding of human reproductive physiology.

Revolt against the conspiracy of silence was under way before 1900, involving such a notable figure as Havelock Ellis, whose series of volumes on the psychology of sex was suppressed by an English court that pronounced it a pretext for the sale of filth. But like a snowball rolling downhill during a thaw, the protest grew by what it touched. The American artist James Montgomery Flagg expressed the clash of ideas when he ended one of his rhymes with the line "To the pure almost everything's rotten".

Especially influential were two small volumes by a distinguished British paleobotanist, Dr. Marie Stopes, that were widely read in the early 1920's. Low birth rates had already developed in France and the Scandinavian countries, but the abrupt trend in this direction among the British after 1923 was no less than amazing; from a woman scientist came what was needed for a phase change, like the crystal added to a supersaturated solution at a critical time.

In France, unlike the United States with its almost pathological fear of a landed class, the law permitted land to be kept within a family. Long before the day of modern controls, the canny French peasant saw the folly of too many claimants on his modest patrimony. So effective was his limitation of family size that our then former president Theodore Roosevelt in Olympian mood chided the French for "race suicide".

Scandinavians are not only literate but also, like the Lowlanders of western Europe, acutely aware of their limited living space, which they must use with skill and wisdom or else lose the advantages of enlightened civilization. And unlike some portions of the earth where folkways and their supporting sanctions are serious obstacles to population control, the Scandinavian people are relatively free from such pressure. Without arguing against the skeptic who sees virtues solely as products of grim necessity, it is fair to say that in Scandinavia the stage was set for a new ethos of profound importance to the future of mankind: *Children are too important to be biological accidents; every child should be entitled to a welcome and to the chance for a good life.*

So important is the relation of human numbers to space and its content that, by 1950, a "sea change" took place among spokesmen for conservation of natural resources. These environmentalists, long concerned with the depletion of materials and energy, and later with the disruption of natural processes such as soil, water, and mineral economy, began to say that no measures could be effective unless the increase of human numbers could be arrested.

Certainly a sheer physical principle is involved here. When dynamic particles are confined, whether they be molecules of gas in a flask or human beings on a land base, their degree of freedom, or mean free path, decreases as their numbers rise. And when energy is added to the system, either by heating the flask or by supplying gasoline to an internal-combustion engine on wheels, the numbers of collisions increase.

Yet the relation of human numbers to space, air, water, food sources, fiber, or minerals is not a simple, straight-forward (i.e., linear) one. The people of the United States did vast damage to its vegetation, animal life, native populations, soils, and streams before 1900 when their numbers were less than half what they are now. Some of the best and most frugal resource management is to be found in Denmark with more than three hundred inhabitants per square mile, in Switzerland with over more than four hundred, and in the Netherlands, cradle of modern agriculture, with more than eight hundred, about ten times the concentration of the United States.

As a further complication, it could be possible at some stages in the development of a society for an increase in numbers to favor better treatment of the environment and its resources by providing sufficient labor to ensure good husbandry. But this would depend entirely upon the values of the society in question and their power to control community behavior. A good example comes from the religious groups that settled in Pennsylvania, Maryland, and Delaware whose life style, supported by large families, made it possible for farms to remain as productive as ever. But even here the process has its limits; trouble has been avoided by the ancient biological practice of swarming, not only into the midcontinent, but even across national boundaries.

The mere opening up of land and its facilities to

common use is no solution, as the biologist-philosopher Garrett Hardin has shown, for the very natural wish of each individual to use his privilege to the limit ripples out into encroachment, scarcely visible at first, on the others until the cumulative effect is disaster for all. Vast areas of overgrazed range in North America, goat pasture in the Mediterranean area, and the spreading desert margins in Africa and elsewhere speak eloquently of this tragedy of the commons.

Control there must be if the human race is not to wear out its welcome on what has been a hospitable planet, amazingly long in becoming habitable for such an animal as man. Without the grasses and other flowering plants and the quadrupeds they support and the materials they furnish, it is hard to imagine human survival. Yet these necessities date their beginnings only from something like one-fiftieth of earth time, though long before the advent of *Homo sapiens*.

Obviously, control cannot be left to those intent only on preserving power and privilege. Nor can it be shaped in ignorance of and indifference to perspective on the human adventure. Thus far, modern technology has tended, by emphasizing its benefits to some parts of human society, to increase the pressure towards equalizing those benefits everywhere, not always with a perceptive eye toward their merits. The process of homogenization distributes not only good but hazard; the latter includes increasing vulnerability to the slightest failure as the system of technology becomes more elaborate.

Brown, Bonner, and Weir in their book *The Next Hundred Years* make this clear, pointing out that if modern technology should collapse, society would revert to an agrarian economy, whose relative stability has long since been demonstrated. Yet these authors believe that scientific technology does have the possibility of salvaging the human adventure—provided that man can learn to live with man. Two of these three authors are distinguished exponents of modern experimental science, the third a student of the human mind. Making due allowance for context, we are reminded of the land ethic of the naturalist Aldo Leopold: the obligation of man to learn to live with nature. As long as the thoughtful and informed place their faith not in tricks, but in human self-discipline, hope is not dead.

Interlude

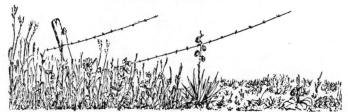

So MUCH HAS HAPPENED with respect to humanity and its environment since the original edition of *Deserts on the March* appeared in 1935 that the first fifteen chapters have been largely rewritten. Unless, however, the reader can share an appreciation of the extent and direction of change, much of the point of this book and its revision will be lost. To that end the next two chapters are reversed in position but reprinted almost as they first appeared, inviting the reader to engage in a bit of vicarious historical experience. He can compare what he reads there with what he observes today and form his own judgment as to whether mankind in general and our own nation in particular are moving toward or away from a sounder and more enduring relationship with the world that has sustained our kind.

Since these chapters were written, the moon has been visited and its other side viewed. Inventions then new within memory have become so commonplace that most of us cannot imagine a world without them. Government has developed massive support for science.

Ecology has become a household word, and the British Empire has dissolved. Scientific societies have been added to the list of organizations concerned with professional and ethical qualifications. The climatic cycle exemplified in the closing paragraphs of the next chapter continues, with drought returning in the early 1950's.

In 1935 prohibition had been ended only two years, World War I sixteen years—only half the time that has elapsed since World War II, the second of these tragic struggles. Agribusiness, starting out innocently enough as an effort to increase markets in industry for farm products, has become an attempt to simulate mechanized mass production not in fabrication, but in growing biological materials upon the extremely complex and fragile system known as soil. Farms have become fewer and larger, with huge budgets for equipment and supplies. The traditional combination of plant and animal industry, operated by resident families, is disappearing, with uncertain if not predictably injurious effects on soil structure. Land ownership has become prohibitive except for those who inherit or are wealthy enough to purchase it. The reckless waste of oil and other minerals, cited on page 240, has come to its predictable sequel, as have the fears of those concerned at having government costs consistently far exceed income. For this as yet we seem to have no better formula than the ancient kingly one of debasing currency or repudiating debt.

Without question the refinement of technology through science has brought vast increases in health, comfort, and convenience to many persons, both in the home and place of employment. But it has proceeded on the unspoken assumption that the supplies of fluid fossil fuel will last indefinitely, or at worst can be replaced by applying science.

A railroad guide published as recently as 1937 shows well over two thousand cities and towns served by one major railroad system. Today travel and transport for most of these places depend upon internal combustion engines on wheels or wings, fueled by petroleum whose price today is roughly six times that of 1937.

A major segment of American industry is geared to produce a continuing flow of automotive units, to encourage their sale, and provide the necessary services. Land use, both rural and urban, is being frozen into patterns that assume continuing availability of private, fossil-fuel-powered transportation along with flying machines for greater distances. In many places it is no longer possible to obtain food and other necessities by foot travel, while the use of buses for longer distances lacks the relative comfort afforded by rail.

Amazing progress in petrochemistry has developed textiles, plastics, drugs, dyes, and even foods to replace plant and animal products, although emphasis on synthetic foods is stilled at present for obvious reasons. The net effect has been to disrupt the traditional combination of plant, animal, and human industry on the land. This pattern has proved not only its viability throughout centuries of change but also its necessity for maintaining good soil conditions. Its failures, however many, have followed a loss of status and respect for the actual husbandman, along with increasing pressure for financial return regardless of consequences.

Perhaps most sinister of all present conditions is the

fatalism that regards war as inevitable along with the unquestioning promotion by both capital and labor of the manufacture and sale of munitions. The average human being in no country, including our own, has any enthusiasm for killing or being killed. The growing practice of high officials to specify a particular power as a potential enemy is, like the behavior in Congress in the decades before the Civil War, an excellent way to prevent the achievement of peace.

Meanwhile, famines continue to occur and deserts to spread, generally where population is otherwise unchecked. Estimates of deaths through starvation are appalling, running into more than one hundred thousand in the region south of the Sahara between 1973 and 1976. Here the desert margin has been advancing at an average rate of three miles a year, while in northern Africa a million acres of crop and grazing land are lost to desert in four years. With the exception of Europe, desert is still expanding on every major continent.

In the United States, dust storms still accompany years of drought when spring winds strike plowed land. And not only in actual desert regions, but also in those less arid, irrigation continues on borrowed time as it pumps up water from underground reservoirs where recharge, if it exists at all, falls far short of use.

Again, however well our convenience is being served by the million or so acres taken out of production annually for roads, industrial and urban development, and the like in the United States, it is important to note that much of this conversion is irreversible in any practical sense. Worse, it often disregards best future use, devouring the most fertile land as readily as the less productive.

Most ominous for the environment, and therefore for ourselves, is the same kind of phenomenon that led to the disappearance of gigantic reptiles in the past. Like dinosaur bodies, our vast political, military, and industrial units have grown faster than the means of coordination and control.

Again, the reader is reminded that the next two chapters reflect conditions as they were in 1935 and is asked to consider, while reading them, what has changed since then.

16

Where is the Sun?

SCIENCE HAS THE POWER to illuminate, but not to solve, the deeper problems of mankind. For always after knowledge come choice and action, both of them intensely personal. Science is like those services that supply a battle commander with information of the enemy and technical advice, but cannot relieve him of the task of decision. For this last, his own judgment, be it rational or intuitive, perhaps even the toss of a coin, must serve. The inescapable character of this crisis is expressed in the old military maxim that action on an inferior decision is far better than no action at all.

The preceding pages were distilled from some years of observation; the more than two-score years since they were written have served only to confirm their essential message of the dangerous unbalance between mankind and his surroundings. In the following chapter we shall suggest that the institution of private ownership, that is, a responsible stake in the game, deserves a genuine trial before it is discarded as a means to achieve the order so essential to welfare or even survival. But we must bear in mind that science as such affords us no sanctions; it can inform but not decide.

The automobile, X-ray, and the conscious use of vitamins are so recent that most of us cannot fail to realize the difference in life that science and technology have brought about. Advertising agencies find themselves obliged to borrow the trappings and the jargon of science, if not its solid truths, in order to inspire confidence in their displays. The microscopes they picture may be out of joint and blind, and the "noted authorities" they quote may be cheap mercenaries, but it all seems to help sell the commodities. Modern science began with the wizards and necromancers of the Middle Ages, and is still invested with the awe that surrounded them. Even the scientist entering the laboratory of a specialist in another field is obliged to pinch himself now and then.

Indeed, the faith that reposes in the human mind for the power of science to work miracles is at once touching and dangerous. It is a curious destiny to befall the branch of knowledge that has done most to free mankind from its cruder superstitions. And because the faith is not groundless, it becomes very important for everyone to understand just how far it should go. There are probably things science cannot do, as well as many that it can. The scientist himself has to work in a very human way, as those who look to him for help should appreciate better than they seem to.

If science is magic, then magic is a very simple thing. The man of science examines the thing in which he is interested thoroughly and often, until he is completely sure about it, and then invites his fellows to do the same thing, to be doubly sure. Touch, taste, hearing, and smell, as well as sight, all aid him in his examination. If he encounters something that, like the other side of the moon, is beyond the reach of his senses, he has to do like the rest of us: guess about it. His only advantage in guessing is that he probably is better acquainted with the part of the moon that all can see.

Sometimes, but by no means always, the man of science can control the thing he is examining and thus learn its behavior under special circumstances. This he calls an experiment. Often he is obliged to take apart whatever it is he is learning about. This he calls analysis. If the thing is fairly simple, he can put it back together, as a child might do with a block puzzle. The chemist often does this in synthesizing compounds. But if the thing is very complex, like a living animal or plant, the scientist is more helpless about putting it back together than a six-year-old with the kitchen clock, which means that each still has a great deal to learn about the thing whose parts are strewn about him.

What is true of the simplest plant or animal is infinitely truer of man and of society. The delicate interplay of motives, the clash of desires, and the seemingly spontaneous growth of culture patterns, so like the irresistible, rhythmic process of birth—all of these involve matters that defy the relatively simple language of science, even when science affixes names to them!

This is not to say that the study of human behavior or of society is a futile occupation. On the contrary, it is highly important. But it does suggest that social control, unlike chemical manipulation, is still, and may always remain, far from being a matter of formulas and routine. Thus, when the scientist has suggestions to offer for the good of mankind, the business of getting them adopted ceases to be a scientific matter and enters the realm of art.

The artist who undertakes to manipulate society—often, it is true, somewhat like a farmer boy manipulating a runaway calf—is the politician. The medium in which he works is that most evasive, at once most pliable, resilient, and difficult of all—human nature. It does not follow that there is anything essentially ignoble in his task, as many so wrongly believe. Certainly he has the courage to attempt a manly enterprise. And if he fails, or if the technique he is obliged to use is not always edifying, perhaps it is less his fault than that of those who should stand at his shoulder with their counsel, their scrutiny, and their support.

In this modern world, where neither height nor depth allows us to escape from conditions brought about by science and technology, no group of citizens is under heavier obligation to assist the man of politics than are the scientists. Yet they cannot, as some fondly suppose, and certainly should not, take over his task. How then, can they best serve?

One does not need to read the tracts of the Chemical Foundation or the utterances of those priests who have gathered themselves around the altar of relativity to know that each discipline considers itself to be particularly potent. Biology, with whose applications we are here concerned, is no exception.

With this—we hope—saving smile at our own earnestness, it may be suggested that, apart from its obvious service to the art of medicine and certain aid in the more immediate problems of agriculture, biology has scarcely tapped the resources that it holds for social use. The fault is on both sides. Society, with a few honorable exceptions, has not seriously underwritten the work of the biologist in the United States. If the young man with the flair for biology wished to become financially secure, he has been obliged to go into medicine, or some vocation in which he could afford science as a hobby. To a limited extent the government service has afforded security of tenure, with retirement pay; but here the scope of research has been necessarily restricted and the better men rapidly drained off into administrative work. There remains the teaching profession. Here the number of posts that give the incumbent the leisure and the facilities needed for creative work are relatively few, and not too often recruited from less favored places. In these latter, which comprise most of the teaching posts, the really tragic loss to society has come not from overwork and underpay, bad as these are, so much as from isolation, discouragement, and lack of stimulus to the men so isolated.

It would surprise most laymen to know that Russia, Japan, Italy, Germany, and even Great Britain are far ahead of us in their employment of science in the public good. The end of World War I left this country without a superior in the fields of scientific endeavor. The American scientist traveling in Europe was accorded the respect due one who comes from a center of his calling. The old attitude of patronage was gone. But we have not held on to our advantage. When budgets have to be trimmed, the laboratories were the first to suffer. The universities, where, after all, a great deal of the world's best research has been done, have saddled their investigators with routine duties to an extent undreamed of in the Old World. Many of the best have had to accept deanships or directorships in order to receive a wage commensurate with their training and the standard of living expected of them. Even the generous allotment made by government for unemployment relief and placed at the disposal of scientific departments could not be used efficiently, because the trained have not had time to direct the untrained properly. Society, like the individual purchaser, in the long run will get just about what it pays for.

In comparison with some of the newer governments of the Old World, we still have some cause for thanks, it is true. Our men of science are not constrained to bolster up an ideology with their findings. Science, religion, and government still preserve their respective identities. The fountain of knowledge still runs pure at the source, though its flow may be checked by a false sense of values among those who should give it generous support.

On the other hand, success is its own best explanation, and the workers in science are not blameless. Too many have lost the vision, permitting themselves to be swamped in needless routine, or have devoted their energies to angling for relatively well-paid administrative jobs. And among those who have achieved professional distinction by their original work, it is the honorable exception who has taken pains to explain to the man on the street what he is trying to do. Yet the greatest have never been ashamed to do this—Huxley and Faraday lectured to workmen, Timiriazeff to peasants.

There, of course, great investigators who lack the skill to interpret their own work. It is likely, however, that many more lack the desire to do so. Their attitude is that of the inveterate golfer or chess player towards his sport—the game is the affair of no one but himself and others equally competent. But only in the laboratory does the man of science have the right to lock out others. His findings, once made, belong to the world, and his is the charge to make them known. Perhaps if the scientist were given not less of technical training, but a great deal more of liberal training than he usually gets, it would make him more directly useful to the rest of us than he frequently is. Effective publication, no less than investigation, is an obligation that rests upon the man of science.

Indeed, there is much for the scientist to learn if he plays his part in the game. He can hardly afford to let it be thought that his word is for sale, as it too often seems to be in the case of expert witnesses. There is little doubt that if scientific societies set out to change this practice, they could do so. The federal courts have shown the way by securing their own scientific commissions when questions of fact are to be determined instead of allowing each side to hire its own witnesses. Scientific groups sometimes put forth exaggerated claims in order to secure financial support. This can end in only one way—the destruction of confidence. The wealth of the scientist, no less than that of the soldier, is honor.

Finally, if the advice of the individual scientist (and

we speak here principally of biologists) is to be worth following, he must fight off the effects that specialization has on him as an individual, without sacrificing his status as a specialist. He must develop and maintain a catholic viewpoint of his field, which in point of fact he seldom has. The battle for recognition of new disciplines did not end with Pasteur's triumphant vindication of bacteriology. It is still in full swing, as anyone who knows the intimate history of biology in America must admit. Illustrations could be multiplied, but the one most germane to our purpose will suffice.

Since 1900 the United States has moved to the forefront in the study of the relationship of land and animal to their environments. This subject, which is natural history in a new guise, has been called *ecology*. It treats of the relationship between the individual living thing and the atmosphere and soil around it, and, of course, the relationships that exist between and among living things.

To the ecologist, a landscape presents a great deal more than its technical details, such as the names of plants or the physical texture of the soil. Rather, it appears as a totality, with each factor, so far as possible, considered in relation to the others. His work involves analysis, of course, but only as a means to final synthesis and interpretation. *When he enters a forest or a meadow, he sees not merely what is there, but what is happening there.* To him, then, there is afforded a glimpse of continuity, integration, and destiny, which is indispensable to management and control in any real sense.

The preceding chapters have been, speaking broadly, an attempt to interpret the relations and adjustments of man as they appear to the ecologist, with due regard for the many intangibles that enter the human setting. If these chapters have told their story, the importance of ecology to plant and animal industry, and to any program of land utilization, should be obvious.

Notwithstanding, a number of great universities do not recognize this discipline, or pay it lip service at best. What is still more serious, the various state experiment stations, even in predominantly agricultural states, have been very slow to add trained ecologists to their staffs. Happily, the Forest Service early received recruits with adequate ecological training, and their influence has been increasingly apparent. But in Great Britain the ecologists are being consulted at every step in planning the proper utilization of those parts of the commonwealth not yet settled, thus definitely ending the era of haphazard exploitation. There are hopeful, but all too few signs that our own national government realizes the part that ecology must play in a permanent program.

So vast and diverse, however, are the conditions that any program must encounter, that it should gather its strength from every part of the country. Neither the ecologist nor anyone else can render advice without information. His problem, in a peculiar sense, lies "in the grass roots." There are probably few counties in the nation in which, over a period of years, a resident ecologist would not yield heavy returns. Certainly in many counties income from lands now rendered worthless by erosion would have sufficed to support such an adviser, as surely as the people on them could support their family physician before the farms went out beneath their feet.

We take it as a matter of course that a city of fifteen or twenty thousand should employ a chemist, even though many do not. At one time such a suggestion would have seemed absurd. The proposal for a local ecologist is a parallel. There are counties, to be sure, in which the situation could be met, provided that the county agent had at least a measure of ecological training, which today he seldom has. But in general the duties of the county agent are so insistent and immediate that they should be supplemented by, not combined with, those of the ecologist.

While the county agent instructs his patrons in the more practical problems they encounter—handling of livestock and crops, marketing, and farm engineering— the ecologist should devote his energy to study and his thought to the future. Thus would he supplement the work of his colleague and furnish the sustaining background of policy, which, as we have seen, is too often lacking in the daily strain of meeting problems directly.

From the studies of the local ecologist, whether he serves one county or a group of them, must come the data indispensable to state and national planning. And from his discourse with farmers and business men we might expect that local measures would frequently take their start, thus lessening the burden of responsibility upon the higher units of government. Indeed such measures would go toward preventing ultimate, perhaps violent, interference in the lives of citizens from those higher sources.

The ecologist, with all of his professional training,

should be chosen with some regard for his talents as a publicist. People, no less than plants and animals, are a part of his material. He should of necessity have the equipment to work with them, comprehend their problems, and admit them to his own confidence, for unless the general citizenry catch an understanding of the whole scene of which they are part, they will not be fitted to participate in a solution of their own problems. And upon their capacity to do so, if they have been honestly and well informed, are free institutions predicated.

Dust storms obscuring the sun for days at a time were raging when the author began writing *Deserts on the March;* today, as the author concludes what has been to him an interesting adventure in applying science to our living problems, rain is falling and has been falling in the greatest quantity since the weather records began, swelling rivers into murky torrents laden with rich farm soil. Before his eyes in the short space in which he has written, the inevitable turn of climate has swung the cycle away from the menace of drought to the danger of flood. Everywhere about him he has seen the landscape as unprepared to withstand the one extreme as the other, thanks to the unconsidered destruction wrought by our haphazard ways.

Within a few months nature has run the scale of her seemingly inconsistent behavior. With rare perfection she has displayed the limits of her caprice. No plainer or clearer warning can we ask. The fields that last year were parched are now gutted. Yet the few remaining prairies and forests are today alive with the beauty of rich growth, even as last year they stood through the withering drought with sturdy vigor.

Surely it should be clear that the grassland and the forest must be restored and protected to an extent not yet dreamed of, not for reasons of sentiment, but because they represent sources of certain return under all conditions. And to the balance they display must man look for his soundest lessons in the construction of his fields to be buffered against whatever may come.

Only with the works of his hands thus attuned to the compelling frame of soil and climate will destructive change take its proper place as a dim memory of the hideous thing it is.

17

The Way Through

IN ALL THEIR RULE and strictist tie of their order, there
was but this one clause to be observed: *Do what thou
wilt*. With those four words the wise old Benedictine
Dr. Rabelais sums up the law that governed the inhabitants of the beautiful Abbey of Thélème. Free, enlightened, noble of spirit, they dwelt in the midst of
plenty. Untouched by fear or need, no one of their
number could find the occasion of his own good in the
ill of another:

Men that are free, well-born, well-bred, and conversant in
honest companies, have naturally an instinct and spur that
prompteth them unto virtuous actions, and withdraws them
from vice, which is called honor. Those same men, when . . .
brought under and kept down, turn aside from that noble
disposition . . . ; for it is agreeable to the nature of man to
long after things forbidden, and to desire that which is de-
nied. By this liberty they entered into a very laudable emulation, to do all of them what they saw did please one.

Thus wisdom plans the perfect state, and the sensitive artist portrays it. In the older dreams all that remained was for the all-powerful benevolent despot to establish it. In reality, of course, wisdom far too often faltered, art yielded to the fleshpots or to despair, and monarchs were not benevolent. In the modern world there is more to be considered. Science and technology have entered the picture, affecting the course of events at every turn. The masses of mankind must be won before they can be guided, and rare is the man who can win them without recourse to devious means.

The United States of North America today is no man's Utopia. Yet if ever man set out to establish a realm of earthly heavens, he did so on this continent. Until the time of its settlement, the idea that man could progress from a worse state to a better was never taken very seriously, at least not so far as this life was concerned. In the wisdom of the ancient East there was no place for such a notion. Change might occur, of course, but it was the remorseless and hopeless change of a huge wheel on an axis that never moved forward—eternally turning, but never advancing.

North America was settled in the fresh bright morning, after the western world had shaken itself awake from the long, immensely invigorating sleep of the Middle Ages. Hope and unbounded faith, crude and selfish though their personal manifestations might be, were abroad. The New World was, in vision, a better world. It makes no difference what craft or sordid resource individuals may have used to gain their ends, the fact is that the migrant throng had one genuine, vital idea—that of progress.

As each new homestead was established and each new community took form, there was little reverence for the past as a guide. In essence, each new center was to be better than the old, and its children happier and better than their parents. The real motto was not the four words above our coat of arms but the four words of the founder of the Abbey of Thélème: *Do what thou wilt.* Like the Abbey, each new haven was richly dowered. There was enough for all.

With the slate thus clear, rational beings might be expected to have made provision, not alone for an equitable distribution according to immediate needs, but for permanent, just, and stable returns in the time to come. Yet as we have seen, the prizes were snatched with little regard for others, and none for the future. A curious blindness this, in a world that professed to be looking ahead.

This is not to deny that there were sages and dreamers, as well as statesmen, who worked at the shaping of our country. Franklin and Paine saw the evils of concentrated wealth, and the former inveighed endlessly against waste, which has been our besetting sin. But their hands were forced by the magic of those two words: "liberty" and "equality." Liberty became license, and equality chaos, with the disastrous results we have been at some trouble to explain in the preceding chapters.

It has been our task, by repeated and insistent emphasis, to show the unbalance man has produced on this continent. And it should be clear by now that this situation is a very dangerous one. Unbalance creates further unbalance, and destruction of the means of subsistence proceeds at an accelerated pace. The picture we have seen is not one of utilization and adjust-

ment, but rather one of exploitation and waste. We
have seen how vast stretches of natural vegetation
have been looked upon as obstacles to humanity, and
destroyed, when in fact they are not only essential as a
safeguard to the normal occupations of agriculture and
industry, but could have been in themselves an unfail-
ing source of steady, dependable wealth. All of this,
having been brought about under what has appeared
to be a system of private ownership and initiative, is
now being cited as a powerful and convincing argument against the continuation of the system.

It is possible to be resigned to the situation. Destructive change is a part of the normal work of nature. If it
is going on much faster than usual, what of it? Matters
may be depended upon to right themselves ultimately.
The mere fact that white humanity in North America
has tremendously accelerated the natural cycle of erosion and change, and completely altered the pattern of
life on that continent, impoverishing it as a habitat, is
of no more importance in the physical cosmos than the
fact that during the Permian period the earth was cold
and dry. Vast glaciers and perhaps vaster deserts in
that age made it impossible for the rich flora of tree
ferns and their kind to survive. The tougher conifers
weathered through, and still retain possession of the
colder, drier forest regions, augmented elsewhere by
the wealth of flowering plants that have arisen since
and established themselves. If we render barren any
considerable areas, we shall of course be obliged to
vacate them. In that event the very absence of animal
life will afford the necessary condition for their recuperation, however slow. If a large part of the continent

becomes unfit to support a population, that population will abandon it until it can recuperate. Why be so concerned about the morrow?

The trouble with this viewpoint is that people, unlike horses, will not stay in a burning building if they can help it. Nor, in their haste to escape, are they likely to be orderly. Once safely out, they are neither very ceremonious and logical in establishing blame for the fire, nor particularly gentle in their treatment of the culprit, real or supposed, when they lay hands on him.

Easy as it may be to foresee the wrath with which future generations will witness their squandered heritage, there does not appear to be much active resentment over the waste of our natural resources at present. Texas is much less concerned about the future oil supply of the United States than she is lest Oklahoma or Louisiana be permitted to drill faster than she. The man who pays exorbitantly for the wood to build a kitchen shelf views the problem in a very immediate way, as does the farmer who cannot meet his rent or taxes because of depleted soil. There is resentment in plenty, but it is directed at persons and institutions rather than actual causes. (Again, see page 211.)

Democracy and the right of private property are two very different things, yet they are inseparably associated in the American mind. They are, moreover, the two institutions that just now are in grave danger. There are, of course, many things in life that are more important than either, but both are the result of prolonged struggle and development. If they are to be cast aside, it should not be for any idle reason.

Their position in the present instance is especially

weak, because it has been through their perversion that the common wealth has been dissipated for the benefit of the few, rather than conserved for the good of all. If people enjoying the privilege of democracy cannot be entrusted with the stewardship of their resources, what could be more sensible, or more humane, than to place them under a benevolent dictator, as the fascists advocated? Or if the powerful persist in robbing the weak, and exploiting them by political and economic means, why not sweep away the whole mess and turn to communism? These are important questions in many minds today.

If we believe that art and manners are prophetic, there is certainly not much ahead. But it is possible that these manifestations represent the fruit of a period that is passing, rather than the germ of what is to come.

Our modern art, and certainly our behavior, is increasingly episodic. Nothing is acted or depicted as though it were related in time, place, or consequence to anything else. Each moment is regarded as an atomic entity, an independent, final thing in itself. We hide away, not for periods of spiritual ripening and repose, but to be something that we are not. There is chaos bewildering, and sadder than the slow spinning of the web of tragic fate. In the latter, at least there was order, cause, and effect, such as sober experience and intuition insist upon. Yet the very fever to crowd as much into life as we do betrays the significance of passing time and the power of human choice, even in matters of no importance. The modern world with a curious inconsistency lets no man forget that time is the stuff of which life is made.

Were one, from a cool and distant vantage, to devise the symbolism of the era we have been passing through, nothing more perfect could be thought of than our manners, our pictures, or our music. Small wonder then that the two political philosophies that have been most vocal are but different facets of the same principle of order through violence, looking toward eventual calm through perversion of the understanding. Each alike contains the canker of its own destruction, for each involves the suppression of individual personality by means of individual force of leadership. Yet in each is an element of consistency and coherence that we cannot boast.

We have developed through the ostensible encouragement of private initiative and responsibility. The beneficiaries of this system are now fighting with their backs to the wall against governmental paternalism. They do not know, or will not concede, that it is paternalism that has made them. William the Conqueror did not parcel out the Saxon holdings with any more lavish hand to his retainers than our government distributed those of the Indians. The Norman king made his awards on the basis of services rendered, and imposed conditions of tenure and responsibility that eventually made the English landed class acutely conscious of its duty to society. We have passed out titles of ownership without assigning obligations in the same measure, and without regard to the final good of all. The paternalism was that of a weak and indulgent parent, moved by present clamor rather than by any sense of future results. It is the children of just such parents who have to be restrained with violence—at the expense of society—when they become adults.

It is unnecessary to go into the unpleasant evidence of our general lawlessness. In large measure this lawlessness may be attributed to freedom without a strong code of responsibility. Curiously enough, this is combined with a powerful sense of order and organization whenever practical necessity arises, as our industrial and business advances show. If revolution ever comes, it will be the result of the insistence for order, no matter how disorderly the means adopted to secure it.

There is food for serious thought in the fact that the two most vigorous political movements of recent decades have risen to power as a result of this demand. The failure of the old Czarist régime was due less to its severity than to its essential disorder. Anyone who doubts that need only refresh his memory concerning the condition and morale of the Russian fleet in the Japanese War, or look into the equipment and training of the huge armies Russia sent to their doom in World War I. The specter of communism has no terrors for an ordered community.

As for fascism, it is an application of martial law to the body politic when disaster threatens because of confusion. In shipwreck or panic or riot, as in battle, it is idle to talk of individual liberty. Salvation depends upon discipline and command, and in time of disorder, intelligent people insist on both. To believe that this country can defend its institutions from either communism or fascism by persecution and propaganda is simply a piece of blind stupidity. Nor could any mistake be graver than to underestimate the strength, the intelligence or the moral earnestness that motivates those who advocate the destruction of our present institutions. They are not to be stopped by bullets, either of lead or gold. Our only defense is to set our house in order, if we think it worth saving.

We have a philosophy of individual initiative and freedom, confused by emphasis on equality. In our wholly laudable desire to do away with the injustices and privilege of the Old World, we have flown in the face of common sense. Science, as well as practical experience, offers abundant proof that men are far from equal in any measurable respect. By our denial of this patent fact we have a nation in which men are less equal before the law than they are today in England, against whose social order we rose in protest. The strong among us, without wasting any effort to demand the external trappings of power, have proceeded to appropriate that power, at the expense of the common good. And as a result, they are free from much of the acknowledged responsibility that should accompany power. They are like the bootlegger, who makes his profits without being a recognized member of commercial life, as is the legitimate liquor merchant. If the latter did not live up to his obligations, it was not because his identity and place of business were not on record.

It must be admitted that those who have profited most under the present system have not displayed the most brilliant generalship during their periods of greatest power. The unrestricted mergers of great industries, which form a perfect precondition for state socialism, were arranged under conservative recent administrations, as were those breaches of corporate trust that have inflamed the popular mind and ended by making the government the mortgagee of a tremendous amount of property. Stupidity, where it handles its own case, has a fool for a client.

As we look over the story of nature, spreading alive before our eyes today, or narrowing back through the long vista of geological time, we see no toleration for structures or organizations that ill fit the work they have to do. Cumbrous reptile and bizarre cycad are alike swept out of the stream of life. Can we expect better for an unseemly and unbalanced social order?

We have had the institution of private property, supposedly, and it is that institution that is now being assailed from many sides. Is it not possible that the trouble has not been with private ownership as such, but with the fact that it has not seriously and consistently been the rule in this country? The abuses of our forest resources and our grazing lands were not essentially the ruination by individuals of what actually belonged to them. Rather they were the result of the irresponsible exploitation of concessions and leases of the public domain, granted to private enterprise by government under conditions of economic advantage such as no business concern would even consider granting to another. No matter what justification political philosophy may have offered, the bargains made by government were unfair to the public it represented, and have had ultimate consequences even more unfair. To grant the right of pillage and exploitation and then blame the results on a system of private ownership with individual initiative is neither just nor straight thinking.

Somehow the thought will not down that both democracy and private ownership deserve to be tried under better conditions than they have yet enjoyed in this country. It is hardly too much to say that we have never had either, in the strictest and best sense. We have said that the exploitation of forest and range is not a chronicle of legitimate private ownership that failed, but of privileged concession or lease of property that belonged to others, often to the public. Even the system of homesteads by which much of the public domain was settled has hardly been a fair trial of private ownership. The size of the allotments has not always been wisely proportioned to peculiar conditions of climate and use. As we have seen, a family might starve to death in the grazing country on a farm of one square mile, while a quarter or even an eighth of that would mean comparative comfort in the beautiful valley of Virginia.

Large irrigation projects have been developed with a system of payments and fees that made the whole enterprise seem very businesslike. But when, because of market and transportation problems, these projects have failed to prosper, people who would never welsh on a personal debt have pressed the government to postpone or remit the moneys due it. Such examples hardly represent normal instances of private ownership, or offer a clear test of the system.

In many places the system of farm finance has obliged the farmer to water his investment to the same relative scale as those of the ill-scented corporations that have been so frequently and profitably bankrupted. No matter if the deed (secured by heavy mortgage) be in the farmer's name, his luckless venture does not represent what he could do with a fair, clean start at private ownership of his farm, under a reasonable system of taxation.

We are on the eve of a determined movement to increase greatly the property held and administered by government. This will in many cases entail the payment of substantial prices in public money to owners whose original title was cheaply obtained, and who have, in the meantime, depleted the value of their holdings. It is reasonable to expect that the burden of these, as of other costly governmental enterprises, will fall most heavily upon the middle classes, already reeling from the successive blows of war and depression. It is axiomatic that political and social stability rests with the middle classes. There is, of course, the question of whether the innovations will benefit those who pay for them, but there is also the more serious question as to how much more this group can bear, without a disastrous disintegration at the lower fringe. If it becomes impossible for people of good capacity to prosper in our system, it is too much to hope that their talents will not be lent to help destroy that system.

The publicly owned forests and other enterprises in the more stable socialistic countries of Europe certainly afford a tempting instance of what can be done. And our smartly managed army, coast guard, and mint show what government control can do in our own land. Our school system is essentially a socialistic enterprise. To a degree these bright examples are shadowed by the dark facts of bureaucratic jealousies, undoubted waste, and chronic political interference. We shall have to make the choice.

One fact certainly must be granted. It is human nature for a man to take better care of his own property than of another's, provided that he has been trained to do so. People living in rented houses are decidedly slower about shutting the windows in a rainstorm than those who live in their own. There should be no inconsistency in managing private property in such a way that it will yield a steady, dependable return and at the same time serve the permanent interests of society.

Perhaps the most potent argument for government expropriation of property at this time is to be found in the fact that resources are being wasted that the commonwealth requires for its future survival. The direct test as to whether this carries weight is the government's will to step in and condemn properties that now are, and in the future will be, highly profitable. Oil and other mineral resources, so essential to modern civilization and now being squandered with a bestial prodigality, are nationalized upon their discovery in Denmark. Would that be possible here? How about the best of our uncut forests now in private hands? What of the choicest range lands, and, to be thoroughly consistent, good farming lands? These represent assets that are worth conserving for the good of all, and if we may judge by experience, there is little guarantee that they will be conserved. From the standpoint of practical politics no one believes that government will go in seriously for the ownership of such resources, unless we have a revolution. But if there is continued insistence for government ownership, it is most likely to take the form of an anemic policy of accepting white elephants. Thus will the policy of an unwise and indulgent paternalism be preserved—at the cost of dissipated public wealth.

So far as the attainment of democracy goes, it is a problem of many facets, of which we may only touch the few that most concern us here. The worst obstacle is the despair of the individual at ever making his one voice heard among so many millions. Perhaps the remedy lies in placing more, not less, actual responsibility on the smaller units of government, while simplifying the purely political machinery of the larger. The situation was stated well enough by an inscrutable president who said that it was essential for small units to decide matters for themselves, but equally essential that in many matters they all decide to do the same thing!

Not only must the individual feel less futile in the scheme of things than he does at present, but, if democracy is to suceed, it must see that he is better informed. In fact, informing him is one of the best ways to make him feel a little less like a nonentity. Only in slippery business is it necessary to keep others in the dark—"the eye of the evil-doer waiteth for twilight."

Needless to say, in giving people information, the best of all ways is to catch them young. A southern banker tells of passing a farmhouse, neatly kept and obviously prosperous. To his intense surprise he saw ensconced upon the front steps an old duffer who was chronically inefficient, notoriously a failure. Stopping to make inquiry, he learned that the place was being handled by the children, trained in 4-H clubs sponsored by the government. Livestock was in excellent condition and of good breed. The garden was well kept, the cellar well stocked. In every respect the place was self-sustaining, and there was, perhaps for the first time in generations, a growing surplus of cash.

If this were an isolated incident, it would not be worth the telling. But it is typical of the situation in thousands of families. By voluntary associations, sponsored by trained and sympathetic workers, a substitute for the old system of apprenticeship has been introduced into American life. There is perhaps no more hopeful sign on the horizon at present than this, yet in far too many communities resources and encouragement are needed to further such work. If the financial and political energy that goes into propaganda of discontent were turned into 4-H channels, it would do no harm.

But the old must be informed as well as the young. Often this is a gesture of piety, but perhaps not so often as many suppose, particularly if the information is made interesting and turns out to be dependable. It is true that an extension lecture on poultry-raising is likely to get a slim crowd if there is a good medicine show in town the same night; in such case the wise lecturer will adjourn and observe his rival's technique.

The governments have made greater efforts toward informing adults who handle the soil than toward any other single vocational group, except their own employees. It would be erroneous to regard the present crisis as caused by a failure of this system of information. Rather, it is one thing that has mitigated and delayed the crisis.

Necessarily, the first efforts at diffusing popular information were halting and tentative. Many of the earlier bulletins, for example, are not worth the paper they were printed on. Scientific knowledge as applied to agriculture is a very complex matter, constantly growing and still far from its goal. Many of the state agricultural agencies were seriously handicapped by politics in the earlier years, and some still are. Public departments interested in agriculture carry a tremendous load of routine duties and are constantly being called upon to solve emergency problems. One of the best soil specialists in the country, for example, is kept so busy with trivialities that he has little time or energy to devote to thought and experiment along permanently significant lines. Even today there is in many places distinct pressure upon scientific staffs for frequent, rather than excellent, publication.

In consequence a great deal of the information handed out is of the nature of that found in a cookbook— useful enough for preparing a meal, but of no account in the intelligent planning of diet. Even the present campaign to enlighten farmers on soil conservation has been possible, not as a matter of foresight and policy, but because the problem has become so acute in many places that it constitutes a direct emergency.

If a farmer consults his county agent or experiment station today with respect to controlling certain insect pests, the chances are he will be advised to clean up his fence rows, grubbing out the shrubs and burning the weeds. As an immediate solution, this is good. But the end results may be quite otherwise. The fence row is likely to be the last stand of the native plants and animals, including the game birds, which are the farmer's best defense, in the long run, against his insect enemies. Or let us suppose a farmer in the drier grassland province contemplates the growing of wheat. He is counseled with regard to the plowing, planting, and cropping technique peculiar to dry farming, perhaps even warned that he cannot expect frequent crops. But of the cycles of the weather and their inevitable ruinous effects, he must learn by bitter experience.

There is, in other words, too much emphasis on detail, not enough on policy. Counsel and advice might very well deal with the problems of the individual as they extend over space and time. It would be interesting to observe the effect of more emphasis placed upon community welfare and the results in the generations to come. Such methods are, of course, too slow for the apostle of direct action, but they might make needless his plan of throwing democracy into the discard.

With a new generation trained in the efficient stewardship of private property, and the older counseled from a rich fund of growing scientific knowledge, emphasizing constantly the importance of policy and management on a permanent basis, the problem of taxation still remains. Intricate and vexatious, it ramifies into all branches of economic, political, and social life, reaching far beyond the realm of applied biology we have been traversing. Probably its adjustment will never be equitable and harmonious. Too many debatable matters are involved. There is little agreement on principles, and less on practice. Yet it is clear that the first obligation of the citizen is to support himself and his dependents. That alone constitutes the discharge of a duty to the state, and no tax is just that moves a family below the line of self-subsistence. Again, there is in this country a constant tendency to penalize by taxation any initiative shown in the management of property. This operates directly and inevitably against any sound policy of conserving resources in the interest of sustained return.

It would even seem possible for government to encourage good management and effectively discourage
management that works against the public interest.
Taxation is a powerful instrument, for other purposes
besides the raising of revenue. In the past it has too
often operated to discourage good management rather
than bad. The man who improves his property in ways
that can bring him no immediate return is likely to be
reminded of his folly at the next assessment. The man
whose property deteriorates suffers no penalty; indeed,
he often escapes the tax increase that comes to the
better citizen.

Thus does the government at the present time join
with those insistent forces that focus our attention
upon the need for immediate advantage rather than
future good. In this insistence upon present profit at
any cost, rather than in the institution of private ownership as such, arise most of the troubles we have been
considering.

Let us suppose that the policy of taxation could be
altered so that the widest exemptions are granted to
the owner whose management is self-sustaining and
whose property is well conserved. Suppose that stiff
rates were imposed upon those properties whose value
is being dissipated by lack of proper management. The
combination of incentive with penalty is powerful. The
man who behaves like a trustee for the future will
have a more tangible reward than the title of "master
farmer." The landlord who holds an acreage for specu-
lation and allows both tenant and land to run downhill
for lack of intelligent supervision would be brought up
with a jerk.

Such a policy would go far to clear up that confusion

of thought whereby the immediate is considered practical, the future theoretical and, by implication, contemptible. And in the end it could be certainly counted on to increase the national wealth, and by that means, the public income. It is not even too much to hope that tenant farming would be greatly decreased, or else elevated from its present degradation, if the mismanagement that it usually presents were taxed out of existence.

The present demand to tax out of existence fortunes and enterprises beyond a certain size may be wise and necessary, or it may prove disastrous. But it would seem, on the face of things, to be a less immediate need than the use of taxation to encourage conservation.

Finally, the obligation of taxes is perennial, and they must be paid in money. Market conditions may be such that produce from the land cannot be converted into cash. At present no mechanism exists by which taxes, as in primitive life, may be paid in kind. Yet during the periods of depression when there is no market, the government becomes an extensive handler of produce in order to feed the unemployed. So we witness elaborate schemes for beating the devil about the stump, the farmer going into debt to pay his taxes in cash—if he is able to borrow the money; the government in turn using the money to buy and distribute foodstuffs and other essentials. On occasion we might relieve the struggle to invent new devices of government by turning back and learning from those who practiced simpler ways of living.

There is no escape from the fact that such a program as we have pictured, essentially moderate though it be, faces tremendous obstacles in the accomplishment.

But we are far past the time when affairs can be trusted to run upon their own inertia. We have had enough of blind driving.

For the perspective of the newborn, which knows no planes of distance, we must substitute that of the mature, with its sense of continuity and proportion. We are not an insensible people, utterly brutish, concerned solely with today and incapable of thinking about tomorrow. But we need to remind ourselves in our quest for immediate subsistence and wealth that while a bird in the hand is worth two in the bush, birds breed in pairs and nest in bushes.

18
Unfinished Business

AT THIS POINT, the reader is invited to turn back to the opening paragraph of chapter one. This to remind him that *Deserts on the March* was written to explain, from the viewpoint of a naturalist, the disaster to man and land in the semiarid grassland province that culminated in the dust storms of the early 1930's. He will also be reminded at the outset that this disaster, so skillfully dramatized later by John Steinbeck in his *Grapes of Wrath,* was not an isolated or unique and capricious event to be shrugged off passively and fatalistically as an act of God.

Instead, it was the consequence of an adventure in land use that was encouraged by financial and industrial interests and carried out by operators, both tenants and owners, bemused by initial high returns. In the stampede for immediate profit, the quiet warnings of those who knew that years of drought were bound to recur in this marginal region, that there are serious limits to the mass production of biological materials, and that stability in nature is based on diversity were not even heard, let alone heeded.

Under these circumstances the natural reaction is to blame a system of political economy, not without its faults, that is variously known as free or private enterprise based on the profit motive. Yet as our analysis proceeded, it became increasingly clear that the exploitation, damage, and often the ruin of man's environment goes far back in history and is no monoply of our kind of society. If further proof be needed, it comes from the experience of the chief exemplar of an economic and political system set up as a living critique of what its spokesmen call capitalism. It was *after* Nikita Kruschev's visit to the United States that Russia repeated our dust bowl experience. Sadly for him, his country, and perhaps for the rest of the world, he was shown plenty of Iowa, but too little of the region west of the 100th meridian.

The final chapter of the original edition (1935) of this book, reprinted herein as chapter sixteen, without substantial change, suggested as a modest beginning in environmental reform that the government support scientific research more generously, that scientists themselves receive broader training, and that sound ecological counsel be made available to citizens where they live, where environmental problems arise, and where informed action must be taken if the individual is to be respected as a political force. Never to be forgotten is Abraham Lincoln's warning that our nation, then under fire, was being tested as to whether it "or any nation so conceived and so dedicated" could long endure.

By 1947 sufficient progress had been made in applying science to land use and management to encourage adding a chapter with the buoyant title "Deserts in Retreat". Federal legislation had made technical advice available to soil conservation districts, visibly reducing soil erosion and improving land use practices. Organizations of concerned citizens, often aided by corporations and credit agencies such as the Federal Reserve, were educating the public and encouraging governmental action. Planning was becoming a respectable word, while conservation literature was expanding from a short shelf into the virtual library it now is. The air traveler was finding assurance of better husbandry in views of terraced hillsides, strip cropping, grassed runways for water, farm ponds, and shelterbelts of trees in the former dust bowl. Highflying jets had not yet made such study of the terrain difficult.

By 1950 the brilliant success of American science in helping to win a second world war had wakened government to the need for much better than haphazard support of research. The establishment of the National Science Foundation resulted; fortunately, its first emphasis was on encouraging basic or fundamental research. Somewhat later this encouragement was expanded to include environmental studies. In 1957 any lingering doubts about the importance of science vanished with news of the successful launching of the Russian spaceship Sputnik I. The modest initial funding of the NSF expanded rapidly.

But like the course of true love, that of science and its uses seldom runs smoothly. Seeming solutions beget side effects not anticipated. The almost despairing concern with overpopulation that had come to the fore by 1950 was soon overtaken by an issue of higher and more immediate visibility—pollution. Essentially a failure to continue a process that goes on quite well in the absence of human beings, i.e., the efficient reuse of materials, pollution has been more intensified than abated by the way in which production is organized to meet human needs in an age of science.

Years ago, when the demand for amusement was less frantic than it is today, there was printed a collection of blunders, or "bulls" as they were then known; among them was this: "Misfortunes never come singly. The greatest of all possible misfortunes is generally followed by one much greater." This bit of dated humor serves to remind us that, topping all other environmental concerns, we now have an energy crisis. Like depletion, disruption of water, air, and soil economy, population pressures, and pollution, the energy problem is an inseparable part of the environmental syndrome.

Having discovered vast stores of liquid fossil fuel, invented the internal combustion engine, and hitched these to the troika of mass production, corporate profit, and advertising, we have put our nation on wheels. Every aspect of the environmental problem has felt the impact; regional differences and the sense of attachment to and responsibility for *place* has been blurred. In the long run, this may prove to be one of the more serious consequences of our obsession with the automobile. In the absence of disproof, we would do well to heed the truism that we take care of what we cherish. When one place becomes as good as another, it becomes everybody's—that is to say, nobody's—business.

Despite heroic struggles with them, the traditional environmental concerns are still with us. As one evidence of the genuine effort under way, at least forty-nine organizations are now federated and cooperating as the Natural Resources Council of America, to say nothing of hard-working local and international groups. But deserts are not, as we had fondly hoped in 1947, in retreat. Arid lands conferences continue, as do dust storms and droughts. Recently, representatives of many nations met in Nairobi, Kenya, to consider the growing menace of *desertification,* a term that translates into the expansion of deserts as a result of human activity. We can at least find hope in this shift of emphasis from coping with nature to giving nature a chance!

Yet we face a curious paradox. The efforts to inform the public have not been wasted; concern is at an all-time high. The optimism, however, that seemed justified in the late 1940's has been overtaken and smothered by worsening conditions both domestic and foreign. Birth rates continue high where food is scarcest and living conditions worst. Much of the arid southwestern United States is living on borrowed time, sustaining its growing economy by pumping underground water that is not being recharged. This appears to be true in parts of the former dust bowl whose lush appearance is enough to deceive all concerned; meanwhile the barriers of windbreaks so expensively installed are being cut away to give more space for crops and even the natural growth of woodland that has protected the vulnerable stream banks is going.

If we think the growing scarcity and increasing cost of modern fuels is a special kind of curse reserved for the privileged, advanced, industrialized nations, we should consider the lot of those who depend on the chief traditional fuel of mankind. Heavily populated areas in the Orient, Mideast, and Central America have been stripped of trees to furnish firewood and charcoal; so serious is the resulting shortage that efforts to reforest have been frustrated by the theft of planted saplings to be used as fuel.

Where rainfall is generous, streams are being dammed to a degree that threatens their necessary free flow, while fear increases as to adequate future supplies of water for thirsty urban centers. Although there are communities of moderate size where enough rain falls within their limits to supply the annual need, there are no means to store or keep it clean from the dirty air through which it comes and the wastes that drain into it.

Travelers report almost unbreathable air in some of the world's greatest cities abroad—no comfort to those here from whom the sun is sometimes hidden on cloudless days. The pressure to trade favors in legislating for a country so vast and varied as ours continues to cost established farmers in areas of favorable soil and moisture heavy contributions to irrigate dry lands for future competition at per-acre costs that have no reasonable relation to the longtime return.

And while the fiscal and social woes of our cities are only too evident, there are ominous changes, curiously

wearing the disguise of progress, in the rural sector.
Farms have become fewer and larger, while the
number of those who work them has decreased in the
three-quarters of a century since 1900 from half to
about one-tenth of our population while unemployment continues high. Throughout history cities have
come and gone, and poor husbandry has often worn out
its welcome on the land, yet experience continues to
prove the viability of the traditional combination of
plant, animal, and human activity on the same parcel
of land. A veteran of World War I once expressed his
respect for the way in which the farmers of western
Europe moved back onto land still steaming from the
heat of deadly conflict, "as they have been doing for
centuries," to use his words.

What generals may see as a convenient place for
battle, the peasant sees as a place to labor at the production of plants and animals needed to sustain life.
Industrialist, sportsman, speculator, artist, scientist—
name any vocation you like—each sees a different
promise in the same view. Under any form of government the political leader, intent for his part on power,
or whatever reward his function offers, has to work for
some kind of resolution of pressures and possibilities
that seems likely to keep the show going.

No wonder that, since recorded history was made
possible by leisure to reflect and pass thought along,
mankind has been concerned to find priorities that
might ensure or even promote order in its ranks. Today, as numbers and powerful technologies press upon finite resources, revolution is inevitable; indeed,
whether we know it or not, we are in its midst. Guid-
ance must come from wherever we find it, not exclud-

ing the rules of ethical and aesthetic experience. But even these realms of the intangible that have done so much to enrich and give meaning to life must honestly recognize those other rules of experience that we are accustomed to calling scientific principles or laws of nature.

Believing this, and reflecting upon more than two-score years of experience since 1935, it seems that the original purpose of this book can best be served by stating some of these rules, leaving the reader to judge for himself the degree to which they fit with what he sees happening around him. Perhaps he will be surprised to see how often these rules express those of folk wisdom and the common sense so often and at such cost ignored in the past.

We have already noted that human beings, as dynamic particles whose activities must go on within confines (for our space vehicles are but extensions of the conditions needed for survival), are subjected to purely physical conditions. As numbers of these active units increase within a given space, freedom decreases. And as energy is added, there is a further average loss of freedom by each individual.

Again, as variety diminishes, so generally do freedom and stability. Because of identical requirements, the worst *potential* competitor of a species is its own kind. Fields and forests of few or single species are the most vulnerable to epidemic. The effective use of environmental materials and energy depends upon many forms of life, each performing in its own way and supplementing the activities of the others. An entire army of generals has little chance to survive.

The doctrine of organic evolution, so strongly braced

by what we know now of the long course of time and process, makes clear that there must be fitness both of organism and environment if their relationship is to continue. Our adventures in space have thus far not shown that any planet in our solar system—*other than Earth*—provides what is needed to sustain life as we know it. This includes not only water, atmosphere, and chemical materials, but the incoming energy at working temperatures suitable to their economy. And evidence is overwhelming that not only individual forms of life, but also the communities they make up, must fit or go out of existence.

The study of human societies, too long concerned with differences, as medicine was with disease instead of health, has at length become interested in whatever is common to all communities. Reinforcing common sense, research and reflection demonstrate that any group of human beings surviving in a given environment *tends* to shake down into some kind of order among its members and between the group and the resources that sustain it. What results are known as folkways or culture patterns. These include not only techniques and customs, but—most important—beliefs and values that reinforce practice by making it seem the right and only way.

History, traditionally more concerned with recording dissonance rather than harmony, has been largely composed of conflicts within and between cultures. Until science gave us insight into environmental processes, little had been said about failures to meet the inexorable terms these processes impose. Chief among those terms is the requirement that to continue, any process—natural or, to use a classic term, moral—

must move toward a working adjustment, an equilibrium or balance, among the components involved. These include the purely physical, such as energy and materials, and the biological—plants, animals, and man. In the absence of evidence to the contrary, it is a safe dictum that no system that runs counter to this rule can survive.

The basic question raised by environmentalist, conservationist, or ecologist is, of course, whether the human adventure should continue as far into the future as possible. When the answer is yes, this becomes the truly conservative philosophy, for it enjoins responsible husbandry of the environment, possible only in a society that disciplines itself.

The idea of equilibrium is familiar enough to the physical scientist, and it is no accident that the British philosopher, economist, and political scientist John Stuart Mill saw the necessity of an ultimate equilibrium in human society. For among his many interests were chemistry and the dynamics of particles. Technically, a system in balance is known as a steady state. In nature this is exemplified by a self-perpetuating living community receiving energy from the sun, doing its work, and recycling the materials it uses. Never immune to changing conditions, such a system is stable in proportion to its degree of organization, vulnerable in proportion to its dependence upon a single factor. This should be evident enough to anyone who has gone through a power failure in the absence of some kind of standby apparatus.

Like him who hath his quarrel just, discipline with a goal that makes ethical, aesthetic, and scientific sense is thrice armed. There are no easy answers, no shotgun prescriptions for the deeply interwoven problems of human society and the planet whose unique fitness has made possible the origin and survival of mankind. At a time when nations are devoting the major part of their labor, treasure, and resources to ever deadlier means of destroying each other, we could do much worse than dust off the injunction that used to adorn the warnings at railroad crossings: STOP, LOOK, AND LISTEN!

Index

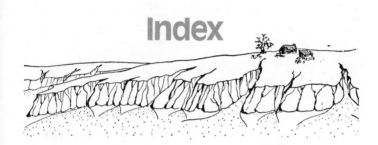